Veneficium

And the serpent is riding over the voice of
the raven, the Other God, to his spouse,
magic, the poison of death, in which Samæl,
the Other God, becomes complete.

TIQQUNEI ZOHAR

Veneficium

Magic, Witchcraft and the Poison Path

SECOND AND REVISED EDITION

Daniel A. Schulke

TRIANGULUM LAMIARUM I

THREE HANDS PRESS

2017

'The Path Envenom'd' was first published in *The Cauldron* No. 126.
'The Matter of Man' was first published in *The Cauldron* No. 123.
'The Spirit Meadow' first appeared in *Clavis Journal* vol. 4, 2016.
A previous version of 'Infernal Vapours' was submitted as an undergraduate paper in 2001 to California Institute of Integral Studies, San Francisco.

Title plate and ornament art by Benjamin A. Vierling.
Cover Art: 'Sacred Heart' Copyright © 2012 by Benjamin A. Vierling.

Interior book design by Joseph Uccello.
Dust jacket and softcover design by Bob Eames.

ISBN 978-1945147203 (softcover)
Manufactured in the United States of America.

threehandspress.com

Table of Contents

INTRODUCTION
The Witch as Poison

In many cultures, sorcery has long been allied to both criminality and heresy. As a series of magical practices in direct opposition to the rituals of orthodox religion, as well as those of 'lawful' folk magic, it has occupied a shadow-realm or state of *otherness* which stands apart from an accepted homeostasis of body and spirit. In Europe, the antinomian nature of such liminal and outlaw magic came to be concentrated in the figure of the witch, whose disturbing rites were expressed in their most complex form in the primeval ritual of the Sabbat, as well as her alignment with spirit-familiars, the Dead, and in many cases the Devil himself. The medieval rite of the Witches' Sabbat—in myth, fantasy, dream, and by midnight in the Meadow of the Goat itself—contains a potent admixture of many religious influences; in addition to the well-known elements of the inversion of Christian ritual, various scholars have identified non-Christian animist and local folk-beliefs, as well as elements of Judaism and classical religions.

Outside the context of the rite of the Sabbat, the aim of the witch as a practitioner of sorcery focused largely on the acquisition of power from supernatural sources for varying uses through spells and other forms of magic. These included divination, congress with the dead, warfare, exorcisms, protection from evil spirits, fulfillment of sexual desires, influencing human or livestock health or illness, locating lost items, procuring love or hate, and attainment of ecstasy.

Poison in various forms was also closely associated with the witch, and used in magical contexts. Similar to witchcraft, the knowledge and use of toxins, from the most ancient of times, has been proscribed by law and religious decree. Like the witch, whose presence in communities was regarded as insidious and polluting, the presence of poison is often undetectable, lurking unseen until it has exacted a toll in victims. This shared countenance of the grotesque inspires fear and awe, and has animated the violent aversion to both the witch and poison over time.

Like the nubile guise of witch, poison may also be alluring, like the intoxicating flower of *Brugmansia*, whose scent may provoke dreams and arousal, but whose flesh precipitates madness. In this guise, both witch and poison serve to arouse and allure, even if doing so with profound unease.

Aside from the dominion of lethality, the magical use of poison may also give rise to excitement, confusion, frenzy, terror, dreams, inspiration, visions, distortion, numbness, pleasure, and healing. All of these components are also to be found, in greater or lesser measure, in the rite of the High Sabbat itself; their correct tessellation gives rise to the *congressus*, a revelatory state of experience beyond the Self. Their superimposition and intersection yields a state of magical embodiment whereby the witch and the poison become co-identified.

Veneficium concerns poisons as a theoretical and practical component of the witchcraft of history, as well as their echoes in modern occult practices, particularly those known as the Sabbatic Current. It is the first of three kindred books treating these practices, taking as their focus various aspects of witch's art. Emerging from occult research ranging from the mid 1990s to the present, its first edition of 2012 gathered together and re-contextualized several essays on this subject that had been published in various places, as well as including several new meditations on the subject. I offer this work to the seeker who would contemplate an occult toxicology, and to exalt both poison and witch as one.

Preface to the First Edition

In 1994, in consideration of seven years' formal study of ritual magic and ethnopharmacology, I began to assemble a book of experiments and operations in the arena of psychotropic sorcery entitled *Hypnotikon*. Its pages held precise *materia magica* and descriptions of obscure psychoactive sacraments, records of the magical practice and expansion of the sensorium into realms beyond my previous knowledge. Also present were detailed reports of ritual operations, performed in solitude, with lodges of varying sizes and magical aims, and also as part of a dyad with a single partner. In some instances I was also allowed access to closed societies as a guest or participant observer, or permitted to record material from informants possessing unique and specialist knowledge.

Alongside this pursuit of the mysticism botanical, my practical studies of herbalism, folk magic, and agriculture complimented this magical tutelage. A deeper passage into the endarkened bower of the wildwood was also a source of direct understanding. Time spent at length in the wilderness served as a clarifying lens for beholding emergent patterning of the work, and as a 'divine emptiness' apart from the impositions of humanity. This vastness was, nonetheless, an Otherworld densely populated with allies, enemies and tricksters, oppressive spirits, voices, storms, telluric powers, trials, injury, death, revelations and above all the ethereal majesty of the eldritch race of færie. It was largely through these solitary encounters that *Hypnotikon* assumed clarity and purpose, and grew into a living and dynamic form. However, the primary motivation for compiling this work was that it serve as a 'primary resource' for my own spiritual development, a record of learning, its essential nature being private. The thesis of *Veneficium*, being a small but potent cup drawn from the vernal well of *Hypnotikon*, is thus a convergence of perspectives I have had for some time: that of the herbalist and the natural magician in congress with poison.

Poison is a glyph for magical power itself: complex, concentrated, liberated in the hands of the elect, and disastrous in the hands of the fool. Its very nature is transmutative, changing all it touches, the maker and breaker of laws, policies, and epidemiological systems. It encompasses what industrialized society conceives as 'drugs', being both the therapeutic agents of medicine, as well as the sanctioned or banned ministers of pleasure and consciousness-alteration. Poison also embraces the toxin: the destroyer and mutilator of flesh, whether weaponized in the arsenal of the farmer or soldier, or lurking in the environment, the scorpion's engorged tail or the industrial pollutant silently pervading the water table. All of these toxicities, for good and for ill, penetrate magic and sorcery on manifold levels, and imply a spiritual discipline of magical toxicology, which in recent times has been called 'The Poison Path.'[1]

Such a path necessarily implies a confrontation with power, but also its effigies, and how these manifest according to the skill of the practitioner. Strewn with thorns, cobbles, and false diversions, the Poison Path offers its own retinue of spirit-guardians to slay, or liberate, the charlatan. Poor judgment, irresponsibility, delusion and ego serve as the great enemies, and in any encounter with poison they must be transmuted or bound long in advance.

1 Dale Pendell, *Pharmako/Poeia, Pharmako/Dynamis, Pharmako/Gnosis*. Similar terminology has been present in 19th and 20th century witch-cults, namely the term 'The Way of Poison.'

The intersection of magic, mysticism, and poison has naturally assumed some concretized and limited shapes in our present era, mostly negative. Perhaps the most important of these has, without sufficient self-examination, been called "drug culture", presented as a legacy of the cultural revolutions of the 1960s. In the 1990s the culture began to publicly redefine itself as 'entheogenic', stressing ethnology, human-plant connections, science, religion, art and pop culture. Typical of this exotery is an insufficiently explored dualism which has haunted the approach of the seeker, that which divides the usage of drugs into 'sacred' or 'profane' purposes. At one side of the ideological bifurcation one encounters the hardened arena of the mystic, devotee, or magician whose approach is one of idealized sacrality in the face of the Gods of Poison. In this stance, of which there are myriad variations, an implicit morality suggests (or preaches) that the legitimate use of consciousness-expanding drugs lies in the realm of the spirit. Common to this approach is the idea of mindfulness, purpose, and direction, as well as the usage as adjunctive to other, less chemical, approaches. At the other side, which might be called the arena of the libertine, is the idea that drugs are to be used for pleasure, socializing, and the pursuits of the body. To some operating within this arena, the 'spirit approach' is a waste of prime vintage which otherwise might be enjoyed by the celebrant, an unnecessary imposition of the mental construct of spirit upon the body.

Both approaches present us with unique teachings, and examples of the uses and abuses of these stances. The Rites of Dionysus, which transect these philosophical zones, remind us that the experience common to both is the generation of ecstasy, of inducing a state of being 'beyond oneself.' Witchcraft, ever transfigured by the magic of need, makes use of both 'sacred' and 'profane' venefic states without adopting these socially-derived boundaries, and far more.

In consideration of the vast and ancient vaults of power poison encompasses, both 'sacred' and 'profane' positions are ultimately moral or religious ones, and in particularly pointed examples, can be considered an unconscious theological debate about the nature of God. Neither approach considers, for example, the use of poison for achieving work or maximizing productivity (as with stimulants or neurohypophysial hormones), mental coercion of others (as with the Haitian *zambi* powder, or sodium pentothal), infliction of pain or physical harassment (as with *Toxicodendron diversilobum*). To this we may add poison's use for destroying malignancy of the flesh (as with anti-cancer compounds such as vincristine derived from *Catharanthus roseus*), building immunity, inducing insanity or indeed treating it. Outside this polarizing rubric also lies the usage of poison as a weapon, either in the arsenals of man or the armies of Nature. All of these approaches proffer a spiritual dimension that may lead to great insights, wisdom, and evolution of the soul. The alternate, shadow-way is also ever-present, which presents the danger of damage, mutilation, and death.

There are also what I have called the philosophical poisons or Astral Venom: the energies, entities or states of being which, though not primarily induced chemically, are nonetheless toxic to mind, spirit and body. As an incarnate being, the first poison is often the attempted destruction of the soul. We may also encounter such toxins in the soul of the curse, unrequited love, an abusive relationship, unexamined motive, and the blight of political ideology. In fact it is a truism that the world, as much as it presents a garden of earthly delights, also presents a concentrated and many-tiered stew of poisons.

Each of these perspectives serves not only to expand awareness of what poison is, but how it functions in the arena of power. In doing so, certain magical approaches and their corpora of knowledge are implied for the practitioner.

THE PATH ENVENOM'D

THE WRITHING HOST OF WORM AND SNAKE; THE VENoms of spider, scorpion, and toad; and the corrupt seeds and nectars of curse-worts all comprise the mythic pharmacopoeia of the witch. Such are the constituents of the cauldron in its sinistral guise, which, together with the hearth-fire itself, seethe in the very midst of the Sabbat's Round.

Throughout time, mundane distortions of this vessel have assumed the grotesques of momentary fancy. In the invective of the religious inquisitor, it is the stew-pot to render human fat and reduce, by depraved ritual, all Earth's accursed to a potent hell-broth. Through the occluded lens of the archetypal, it is become an intellectually palatable 'cauldron of transformation' or womb of feminine mystery. Yet to the witch, whose commerce with the writhing host is bound by the sorcery of spirit-liaison, it is, and has always been, the Black Well of Execration, source of *veneficium*, the magic of poisons, for the power of both good and ill.

Poison, like *witch,* is a word burdened with problematic associations. In vulgar parlance it has come to indicate an agent of destruction, be it of spirit, mind or body, whose inevitable bequests are wounding and death. However, ancient definitions often implied a poison's healing nature as well as its capacity for harm. The Latin *veneficium* can be interpreted as drug, poison, or magic; the old Greek *pharmakon* could equally indicate a poison or a cure.[1] These paradoxical definitions of toxins, wherein powers of both destruction and healing are present, suggest a certain lost wisdom in the magical pharmacology of antiquity, and are aligned with the folk-wisdom of the witch.

To the sorcerer, poison is the ingress of external power initiating crisis. The fruits of its temporal field are the product not only of dosage, but also such factors as route of administration, strength of the physical vessel, purity of the deed of Arte, and the favor of the spirits presiding. As with all destabilizing power, the repercussions can create or destroy; liberate or repress, illuminate or obscure. As alchemical adepts well knew, poison is the point of first beginnings from which all must arise; it is power, both in its raw state, and in all its potentials for transmutation. As Paracelsus noted, it is both omnipresent and absent in Nature.

Self-poisoning for the attainment of mystical knowledge, ecstasy, and congress with spirits, we call 'The Poison Path.'[2] This designation separates the mystical endeavor of Transmutation from the vulgar dross of hedonism or criminal activity. Ours, therefore, is an art of subtle discrimination, of observation and caution. Gnosis of the Poison Path arises not from the first matter of its toxin, nor its mundane somatic effects, but in its Trans-

1 *Viscum,* or Mistletoe, a plant known for its poisonous as well as its curing powers, ultimately derives from the ancient Proto Indo-European roots *Weis* 'to melt away, flow'; Sanskrit *Visam*—poison; Avestan *Vish*—poison.

2 The most thorough philosophical exposition of this endarkened by-way are the *Pharmako* volumes by Dale Pendell, which regard sacred poisoning as one of the most ancient forms of the Art Magical, and Eve as its patron saint *par excellence.*

mutation via the Art Magical to serve the path of the seeker. This Art is thus the holy dominion of Shiva Vishpan, or Shiva the Poison-drinker, emblematized imbibing venoms from a conch, and whose blue skin-color resulted from its ingestion.[3] In Islamic lore, it is the domain of the *uphir*, or Hell's Physician, possessing the secret lore of medicines and dead bodies; akin to the Bulgarian *vapir* or vampire and the Turkish *upir* or 'sorcerer.'

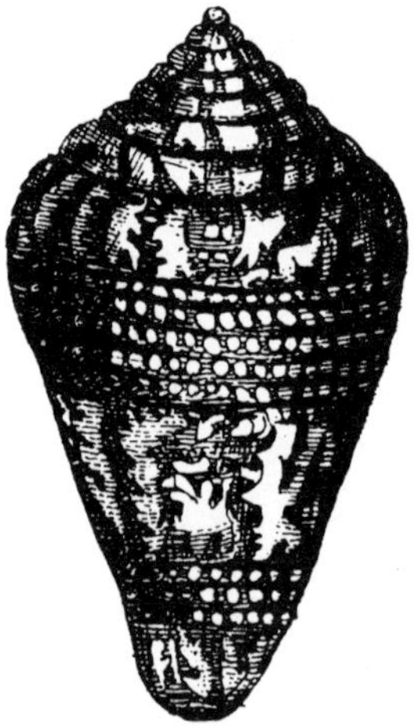

I. CONUS CEDONULLI, *a purveyor of the conotoxin venom complex, whose symptoms include muscle paralysis, distortion of vision, respiratory failure, and death. Antivenoms are elusive.*

Among the great tributaries of ancient magical knowledge flowing to the present, certain forms of Traditional Witchcraft reckon the fruit of Eden's Tree as the dispensation of Samael, or Lucifer, unto mortal kind. As a so-called 'fallen angel', this intelligence is thus the herald of witchdom's dawn, emissary of light to transgress the first tenebration *thou shalt not eat*. Diverse ancient cosmogonic narratives also relay angelic beings as bringers of forbidden power; many are also considered gods of poison. In one ancient pseudepigraphic text, Eve relates that the fruit of the Tree was sprinkled with the Serpent's poison, which she then ate after giving an oath to the angel.[4] Here is found the Poisoning Art in its spiritual emergence, conterminous with other primeval deeds of witchcraft: congress between angel and human,

3 This azure tint brings to mind certain cone shells of the family Conidae, common in the seas of the South Pacific, which possess a dart-like proboscis capable of injecting a painful and potentially lethal neurotoxin. One of its poisoning symptoms in humans is cyanosis, or blue skin coloration.

4 *Life of Adam and Eve*, 19:3.

transgression against the demiurge, self-liberation, and the assumption of New Flesh.[5] Commerce between the Serpent and Eve thus inaugurates a pattern of initiatic process which continues throughout magical time. That these beings were male, and their first human initiates women, invokes certain arcana of sexual sorcery.

An essential feature of Traditional Witchcraft is rural life and its allied corpus of tradition, lore, and custom. It thus serves as a repository of the practical teachings of farm and countryside, true to its source, despite centuries of contact with urban influences. This peculiarity is not ossified—it is alive and actively engaged by those whose trades are shaped by the land and its laws. Part of this granary of knowledge concerns the immediacy of death in daily life, whether it be the killing of livestock, hunting, the destruction of vermin, or the threat of human death from varied causes. Rural knowledge readily accepted this aspect of its world; from the storehouse of country wisdom comes our great legacy of common plant names, many of which carry the warnings of their venomous nature. For example, the suffix '-bane' appears in the common names of numerous botanical poisons such as cowbane (*Cicuta* spp.), wolfsbane (*Aconitum* spp.), dogbane (*Apocynum* spp.) henbane (*Hyoscyamus* spp.), and baneberry (*Actaea* spp.).

The skilled ritual use of hellebore, fly agaric, as well as several toxic plants of the nightshade family, has been present in several closed circles of Traditional Craft to which I belong.[5] Where custodianship of these arts is held as an office within covine, the one who stands in this station may bear such names as Verdelet, Hayward, Doctor, or Green Mantle. More often, however, specialization of herb-cunning is held among many members

5 A mystical concept of Sabbatic Witchcraft relating to magical augmentation of the Self, drawing upon ecstatic ritual, atavistic power, and the sorcerous formulation of Will, Desire, and Belief. See Andrew Chumbley, 'Wisdom for the New Flesh' and Schulke, *Lux Hæresis*.

of the circle, according to interest, training, need and proclivity. In addition to practical knowledge, some of these teachings are mystical or poetic, or survive as simple charms, suggestive of an earlier historical context in which knowledge of poisons and counterpoisons was prevalent. However, present alongside such traditional teachings are certain admonitions. The chief of these is that the use of visionary plants is in no wise a substitute for a sober and rigorously-focused magical practice, and that the verity of spirit-congress may easily be contaminated by the very poison regarded as sacred. Thus, within the Circle of Art, the Magical Poison is but one color on the palette of the Divine Artist.[6]

This is not, however, to dismiss the magical potentials for the visionary sacrament, for such have existed in traditional rituals from the dawn of sorcery. One such example of particular relevance to European magic, and the oneiric magic of Sabbatic Witchcraft in particular, was documented by the nineteenth-century researcher Julius Klaproth. These were the rites of the Ossetians of the northern Caucasus, early modern descendants of the ancient Scythians. Dominant in their religious iconography was the Prophet Elijah, to whom goats were sacrificed in caves for a bountiful harvest and to avert hail. They practiced a rite in which leaves and branches of *Rhododendron caucasicum* were smoldered in sacred caves. The resulting smoke from the plant drew seers into a sleep pregnant with omen-rich dreams.[7]

Other areas of investigation relevant to the witchcraft prac-

6 Knowledge of ritual poisons is also present in other traditional witchcraft groups. Robert Cochrane, past Magister of the Royal Windsor Coven, possessed knowledge in this area. See John of Monmouth's *Genuine Witchcraft is Explained*, and Gavin Semple's 'A Poisoned Chalice', *The Cauldron* No. 114.

7 Carlo Ginzburg, *Ecstasies, Deciphering the Witches' Sabbath* pp. 162–3. Visionary poison is not limited to the heathen, as the serpent-handling Christian Pentecostal sects in rural North America reveal. Interaction with live venomous snakes and drinking strychnine, when they do not result in fatality, often bring religious ecstasy and spirit-contact with God the Father, Christ, or the Holy Spirit.

tices is the so-called Witches' Flying Ointment of legend, an unguent which, despite the mystery surrounding it, has a number of ethnobotanical precedents. For example, the Aztec priests of the god Tezcatlipoca ('Smoking Mirror') compounded an ointment called *Teotlacualli* containing copious amounts of tobacco, the ashes of poisonous insects and spiders, caterpillars, salamanders, vipers and soot. To this was added crushed seeds of ololiolique, (*Turbina corymbosa*), a plant of the morning glory family. According to the writings of the Spanish prelate Acosta, the priests thus smeared "...became cruel in spirit. At night they went alone into obscure caves and boasted that wild and savage beasts feared them because of the strength of their mixture."[8]

Sacred Poison may serve as a visionary, but it may also bestow physical distress as magical catalysis. The wayside herb stinging nettle (*Urtica dioica*) is not commonly thought of as poison, but those whose bare flesh has been excruciated by scores of its stings at once will know a measure of its ferocity. The ancient ritual practice of urtication, whereby ascetic devotees of the Christian cultus mortified their flesh with nettles, probably has sorcerous antecedents far more ancient; the histamine shock to the flesh and corresponding endorphin release, in a focused ritual context, can indeed precipitate ecstatic trance. Ceremonial infliction of pain—in this case blistering irritation—finds corollaries in a number of important rites, such as the Pe'a, the tattooing rituals of Samoan men, which, among other purposes, create a sympathetic simulacrum of the pains of childbirth. Irritant poisons as a source of power are also recognized among the Kawaiisu tribes of California, who esteem red ants as a holy medicine.[9]

8 See also Ruiz de Alarcon, *Treatise on the Heathen Superstitions that Today Live Among the Indians Native to this New Spain*, 1629.

9 Maurice Zigmond, *Kawaiisu Ethnobotany* and *The Supernatural World of the Kawaiisu*. In addition to irritant effects, the live ants, balls of which were swallowed in eagle-down, were also a hypnotic and preserver against evil spirits. The Kawaiisu also practiced walking naked through thickets of Stinging

Apart from the use of Sacred Poison upon the body, poison may manifest as sign or omen. The stings of poisonous flora or fauna may serve as the summoner or patron *daimon* of the Witch, heralding the unfolding of blessing or bane, or the indication of poison as a source of her power. Like the bestial retinue which attends upon the dark children of the serpent, a harrowing illness too may also serve as the toxicon's *ordalium*, and in some cases bring dream or vision. Certain other signs, while not poisonous in the chemical sense, may still emanate foul influences or contagion, such as the evil eye. An extrapolation of this was the medieval belief that the gaze of a menstruating woman could dull mirrors, which gave rise to association with the basilisk; transmission of venomous humors was thus passed to the unsuspecting via the passive receptacle of the eye.[10] Objects or places may also be considered exceptionally malefic, and thereby conspecific with poison; an example is 19° Scorpio, known in Arabic astrology as the "Accursed Degree" and identified with the star Serpentis. In all cases where contact is made with such forces, it is the successful transmutation of poison by the sorcerer which determines its potential for gnosis, thus liberating its true power.

The Poison Path echoes the dual ethos of healing and harming found as a strand of many forms of Traditional Witchcraft. This is echoed in the domains of certain deities. Like the ancient words used for poisons, certain gods of poison also command healing, as with Gula, Sumerian goddess of healing and fertility, also linked with poisons and sorcery. Omolu, the divine plague-doctor of Fon Vodou, wields the power of contagion and virulence, but it is He who is beseeched by the faithful for deliverance from disease. This duality may extend to poisonous

Nettle to gain power, a rite reminiscent of the 'Rite of Making Green' present in one form of English Witchcraft, where power is gained directly from the wilderness (See Chumbley, A. 'The Secret Nature of Ritual').

10 Jacquart and Thomasset, *Sexuality and Medicine in the Middle Ages*, pp. 74–75.

plants themselves: the mandrake, for example, is a plant with a well-documented and ancient attachment to sorcery. As well as its place in enchantment, its drug virtues have long been known, including exhilaration, sedation, anesthesia, and lethal poison. These characteristics, as well as their utilization in diverse kinds of magic, have led to its association with the Devil. Yet, the man-root's diablerie notwithstanding, it was also co-identified with Christ in early Christian theology.[11]

As a young man, part of my spiritual training was in the so-called *curanderismo* or Mexican folk-magic of the southern United States, a syncretism of Native American sorcery, Roman Catholicism, and vestiges of European magic, particularly Spanish witchcraft and the grimoire traditions. Though the name for this art directly invoked curing, its attendant body of lore and magical praxis placed equal emphasis on so-called 'left-hand' spellcraft. By strict custom, the practitioners known to me referred to themselves not as *brujos* (witches), but rather as *curanderos*, or sometimes *charismos*, and *mysticos*. In this respect, my own teacher's self-identification was as a 'white magician' whose magical repertoire also included practices to induce death, disease, torment or other harm, as well as healing. Such rites of cursing were considered 'defensive magic', used against 'black magicians': those who exacted these baneful rites had no compunction calling upon both Christ and the Devil for aid in the same prayer.

This dual magical ethos of helping and harming, and the complex mysticism which reconciled their spiritual polarities, was called 'The Crooked Path' by Andrew D. Chumbley, who identified it as a pervasive feature of folk magic and sorcery.[12] The sinister arcanum of the Opposer,[13] a deific embodiment of trans-

11 Hugo Rahner, *Greek Myths and Christian Mystery*, p. 247.

12 Andrew D. Chumbley, *Qutub* (Xoanon 1995), and "The Crooked Path I and II", *Chaos International*, 1992.

13 Some have posited Set, al-Shaitan, Satan or the Devil as this particular deific form, but within Traditional Witchcraft I am personally acquainted with the lore of Cain and Lilith as the forms of the Opposer.

gressive power, is the force of magical catalysis summoned in the moment which allows the overthrow of Self, and thereby transmutation. To the Crooked Path sorcerer, poison is the power of opposition itself, defined not by its mundane toxicity or its potential damage to the flesh, but by its capacity to facilitate Otherness—the twilit dominion of the Witches' Sabbat where temporal reference is suspended and the world becomes malleable. It is thus the Manifest Trident as opposite arcs of the Poison Path: balsam, poison and the Opposer as the transubstantiator between, the mediating path between Panacaea the all-healer, and Pantrauma, the all-wounder. The Path Envenom'd is thus the current manifest against Nature herself, the self-slaying genius of opposition, and the God Without: all must be turned to serve the Path of the initiate. This is the ethos present in the magical transmutation of Poison.

II. *Bestial archetypes of drunkenness, from Thomas Heywood's* Philocothonista, or, The Drunkard, Opened, Dissected and Anatomised, *1635.*

Of equal concern to the particulars of earthly toxins is spiritual poison and its attendant host of powers, which produce tenebrity of spirit. Like their material counterparts, their entity retains an external allure, but beneath their masks writhe an infectious host. The sorcerer does well to consider the three Buddhist Poisons of greed (*lobha*) hatred (*dosa*) and delusion (*mola*). Whilst these attributes are variously regarded in the world of men, to the sorcerer they are the very keys of falsehood, and impediments to power, and the seeds of self-defeat. Spiritual Poison is transmuted by the sorcerer through the Formulae of Opposition, but also by way of the traditional healing charm, and magical methods to 'earth' poison into a fetishistic substrate.

A curious instance of this is the 'Fear Cup' (*Tāsat al-Tarba*) of Islamic folk-medicine, a brass bowl engraved with prayers to angelic divinities, and having suspended around its border forty small brass 'keys' engraved with prayers. The cup is exposed to the night air, and the dew thus collected is drunk by the afflicted to cure physical symptoms having been struck by fear or supernatural terror.[14] However, where spiritual poisons are concerned, the most important of such venoms are not self-administered, but given of the Fates: disease, heartbreak, the death of a loved one, a stroke of ill fortune. It is not mere coincidence that *Atropa belladonna*, elf-queen of poisons, is named for the Lady of the Shears: those who have truly escryed her black mirror know the proximal liaison betwixt revelation and adumbrated finality.

From the Crooked Path perspective, knowledge of the Poison Path must concern itself equally with Poison's opposite. Such knowledge usually encompasses counterpoison or antidote, the magical hypostasis of which is balsam or nectar. Traditionally this has been the realm of the theriac or mithridate, the ancient preparation containing both toxins and therapeutics, used as a 'hedge' or preventative, against poison. Dioscorides made a confection of viper's flesh called *Sal Viperum*, as an antidote to

14 Abd al-Rahmān Ismāil. *Tibb-Al-Rukka* (*Old Wives' Medicine*).

poison; the snake was roasted with figs, salt, honey and Spikenard.[15] Official Mithridate, containing over 60 ingredients, was compounded and prescribed up until the 19th century. The principles and concerns of the Mithridate and counterpoison continue presently in modern medicine in the form of inoculation and vaccination, powers constantly at play in Nature herself.

The mysteria of Poison and Antidote demand that the herbalist look not only at the healing potentials of poisons, but also at the harming potentials of those worts normally considered innocuous or curative, and all states between. Biology instructs that in plant and animal evolution, venoms arose as defensive boundaries, but also as offensive weapons. This knowledge contains the recurring Arcanum of Opposition we have encountered thus far on the Poison Path.

Progressive doses of poison are ever-larger gates unto the spirit-realm, the last of which is physical death. Not all gates are pleasurable, and many bring great pain.[16] Yet within the ambit of the Art Magical we may posit schemata of gnostic poisoning, occupying a continuum whose polarities are benefic and malefic, healing and harming. Movement along the axis in one direction is progressively innocuous, in the other, progressively lethal. For some poisons, there exists betwixt these two points a state of divine madness, the rapture of *ecstasis* wherein the personal 'I' is transgressed and sorcerous traffic with the spirit-world is facilitated. This continuum is expressed differently with every plant or animal venom, and in every enchantment utilizing it. Strychnine, which is a stimulant in small doses, becomes a fatal nerve toxin at high ones.

A perhaps more germane exemplar lies with the visionary nightshades, a particular group of plants sometimes referred to

15 Edith Grey Wheelwright, *Medicinal Plants and Their History*, p. 81.

16 The myth of the 'peaceful death' using natural animal and plant poisons finds little corollary in practical reality. With the notable exception of opium, most lethal botanical doses produce excruciating and ghastly side effects.

as the 'hexing herbs.' These are plants of temperate and tropical distribution with ancient histories of magical use, and characterized by demoniac visions at shamanic doses. The most important brethren in this clan include henbane, belladonna, thornapple, mandrake, angel's trumpet, and tobacco, which, when used traditionally and at sufficient dosage, is hallucinogenic. Though each of these plants is chemically and magically different, they all correlate to similar gnostic continua. From low to high dose, these nightshade 'gates of poison' are:

1. EXHILARANT
2. APHRODISIAC
3. INEBRIANT
4. STUPEFACIENT
5. PHANTASMAGORIC
6. ANESTHETIC
7. FATAL POISON

Revelation may wait behind every gate, but the doors may also slam shut—proffering naught but agony and chastisement!

Yet, for those who tread the Poison Path, perhaps the best knowledge of antidotes lies not in any enchanted balm or chemist's preparation, but in the balanced and proper devotion to Fear. In its exaltation, Fear is the simultaneous knowledge and respect of those powers which can annihilate us. Its irrational axis, manifest in action, is cowardice and impulse; its flowering is courage and prudence. Robert Cochrane cautioned that the use of the poison sacrament in the hands of the fool was a 'quick way to the underworld of insanity', and this skull and crossbones stands as a waymark on the Path Envenom'd. Other spirit-roads may abide the fool, but not the Dark Lady of Venoms, for mercy is a virtue alien to her dark droplets.

Purity, Contamination, & the Magical Virgin

SORCERY AND TOXINS IN COMBINATION IS A POTENT admixture, distilled over the course of millennia, amalgamating meaning, implication and emotional investment. How a practitioner of the Art Magical defines poison is wholly another matter, and self-conception of poison itself is an essential part of the work of the adept.

Kindred concepts implied by the dual archaic usage of the Greek *pharmakon* (poison and magic) lead us to further consider the distinctions between sorcery and intoxication. The word 'intoxication', whose usage in modern English is almost universally linked to the effects of intentionally ingested drugs or alcohol, was once used in pathology to mean poisoning in general, more often than not connoting harmful physical effects as a result. And yet, the adjective 'intoxicating' has in the present day become entwined with notions of ecstasy, allure, and sensual delight in a poetic sense, if not the *lingua vulgaris*. The word 'toxic', by comparison, has emerged at the fore of chemistry and environmental science to strictly denote a substance of potential physical harm or fatality, while in recent English vernacular it may refer to emotional or institutional opprobrium. Venom, linked etymologically to *veneficium,* implies poisons of animal origin, but may also refer in common parlance to human malediction, in word or deed. The multiplicity of meanings that poison and its associated retinue of concepts encompass may seem contradictory to the linguist, but to the magical practitioner possess an internal consistency suggestive of infinite power.

In European witchcraft, as it came to be shaped through the printed word, the linkage between poison and *maleficium* prompted many attempts at clarification of terms. Johannes Weyer, in his *De Lamiis* of 1577, remarks:

> Venefici *are those who with poisons made from metals, plants, animals, excrements, or mixed bodies, swallowed, rubbed in, or placed where their vapors are breathed bring diseases with the most cruel symptoms, wasting of the body, imbecility of strength, loosening of the joints, and other atrocious sufferings, sometimes prolonged but generally causing speedy death with intolerable pain.*

Weyer, a student of Henry Cornelius Agrippa, here advances a purely materialist explanation of the poisoner-witch. Though unquestionably a man of God, he was ever ready to attribute enchantments and fables of witchcraft to superstition, or an otherwise rational explanation stripped of magical romance. Historians frequently laud him as an early and praiseworthy example of a skeptic counterpoised to the 'witchcraft delusion', but the above position also anticipates the modern discipline of criminology. The position is echoed in the *Tractatus de Sortilegiis* of 1592 the Italian jurist Paolo Grillandi:

> *With these compositions and poisons they can render men impotent and women averse to congress...all these diseases and*

> *deaths and impotence and love philtres produced by these mixtures are not the effect of the ceremonies and observances and abuse of sacraments described above.*[1]

Both definitions may fall within a legal framework, but also, if we consider the predominating currents of the time, that of Natural Magic. This discipline, straddling magic and science, was concerned with harnessing powers emanating from Nature direct, many of which bore the trappings of the miraculous. These cladistics, considered in distinction to sorcery, were anticipated by Ibn Wahshiyya al Nabati in his *Book of Poisons,* circa 850 AD:

> *Poison is something which overpowers and destroys that which is called the life force of an animal. When it overcomes this force the functioning of the organs in the body is disturbed. The liver, stomach, and veins cannot function so that the strength of the heart, liver, brain, arteries, daily warmth and sinews cannot be transported through the body as they were previously. The quality of this condition is the property of death, since, in consequence of these things, it corrupts the breath which gives rest to the body. Then the life of the animal is affected without delay. This is the definition of poison.*

In this important volume we have a thorough formulary of plant, mineral, and beast-derived venoms as gathered by Wahshiyya from diverse ancient and contemporary works, some of which are now lost. Each is accompanied by a painstaking description of its preparation and administration, as well as the ghastly effects each wreaks upon the body. Expanding the book's mandate beyond lethality, a considerable number of antidotes are also given. The *Book of Poisons* resists simple classification, but is

1 On a separate level the definition is of interest due to its mention of impotence-causing sorcery, a subdivision of malefic magical charms which persists today in various forms of traditional magic.

nonetheless considered a classic treatise of Arabian toxicology. Beside the more scientific expositions on the nature of poison and a clearly empirical strand informing the author's approach, we also find sections on magical poisons, such as the preparation of a poisonous drum, whose sound kills those who hear it, or fatal compositions, like the fœtid emanations of the Gorgon, which slay merely upon been glimpsed. In many formulae, despite carrying out a complex compounding procedure, the final ingredient animating the mixture to efficacy—particularly the antidotes—is 'God's help.' These definitions stress the corrupting nature of poison, and despite their more rational approaches to the subject, presuppose a contrasting physical homeostasis.

To my own experience as a magical practitioner, there is an additional criterion worthy of our consideration. Poison, in its characteristic modality, is marked by its origin *beyond*, occupying a station exterior to the adept. Whether descending from the fang of the viper, the grail-cup of the Beloved, or as an astral parasite uncoiling from the abyss, poison contains within it an inception of otherness, or power alien to Self, and thus is an embodied metaphysic of the Sabbatic concept of the Opposer.

Simply defined, the Formula of Opposition describes magical contact of the adept with that which lies beyond, and the liberation of power thus resulting.[2] In its higher mystical forms in Traditional Witchcraft, it is the perfect embodiment of contrary magical principles as trans-mutative power. When conscious and active in the initiate, it has been termed Crooked Path Sorcery, and as an ethos driving magical practice may be found in the most ancient magical texts, up to those of the present day.

Returning to the realm of the healer, the Formula of Opposition is also found in certain protocols of pharmacy. The art of utilizing deadly poison as medical therapeutic is a fundamental feature of medicine, especially oncology and the use of antivenins. Here, the governing script concerns right dosage, in perfect

2 Chumbley, *Qutub*, and *The Dragon-Book of Essex*, passim.

resonance with the iatrochemistry of Paracelsus. The botanical family Apocyanaceæ or dogbanes, is rich in plant exemplars harboring potent medicines, such as the genera *Catharanthus* and *Vinca*, providing alkaloids as Vincristine and Vinblastine, important in chemotherapy for the destruction of neoplasms.[3] However, an important feature of such therapeutics is the selective poisoning of infected tissues. In further consideration of the Poison Path, magical principles of Opposition are also embodied in the form of the antidote, which Arab physicians of old sometimes referred to as 'the remedy which opposes.'

The cartography of poison belongs to realm of the other, and its function as a power of opposition is resonant in some measure with the ancient Anglo-Saxon concept of 'flying venom,' elf-shot and *wyrms*, all contaminants emergent from the Outside. Though different in ontology to our modern pathogenic theory of disease, both share in common an exterior origin of poison and corruption.

This magical relationship between interior-exterior is especially evident in the magical operation of the exorcism, the work of purification or a 'casting out' of malign spirit-power. Indeed, the charm *be far from here, all ye profane* contains within it the designation of malign influence as external from the operator. The corpus of European high magical ceremony found in the grimoires offers a precise anatomy of exorcisms, as it is a corpus particularly rich with them. In addition to its cleansing function, the exorcism may also serve as a protective barrier or prophylactic against astral contaminants. Whether the entities conjured are celestial, infernal, planetary, or the discarnate dead, these technical functions link the exorcism to similar practices in Christian liturgy. In the arena of rural sorcery, the exorcism may take a more direct route in medicinal magic. Folk magic has

3 Nature provides other exemplars in abundance, such as the lectin Vincumin, extracted from Mistletoe (*Viscum album*), and desmoteplase, derived from the saliva of vampire bat species, useful in the prevention of stroke.

long made use of charms against poison, of which several types have been identified. Of note are charms against snakebite, of which one taxonomic category is *Snake Bit Christ Spoke*, allied in essence and ritual technique to charms for easing the damage caused by burns.[4] The number of charms for countering poison and for expelling noxious spirits is vast.

The presence of these concepts in exorcism and magical thought once again exposes strata of magical ontology emphasizing purity and contamination, an essential consideration for the adept of the Poison Path. These draw obvious parallels with the arts of medicine. Foundational medical training begins with hygiene, and the best and most effective magical tutelage is no different. Sterile procedure, meant to contain pathogens and avoid cross-contamination, bears many similarities to magical operations, not only of exorcism, but also divination and the magical circle. Relevant too is recognition of one's own potential to infect, and taking responsibility in containing it. In another example, when administering care to a patient, the portions of the body must be bathed in a specific order, in order to preserve the purity of the cleaning vehicula; similarly, magical operations often follow specific sequences of ritual action identifiable across diverse eras and locales. The adept will recognize this is less a reflection of rote adoption, but rather that certain operations have a progression of power that, by their nature, is both cumulative and effective.

In currents of magic, ideas of purity and contamination are often imported from religions: Islam, Hinduism, Christianity, and Judaism. Similar frameworks occur in certain Native American societies such as the Navajo, where the dead are known as corrupting influences, and a retinue of charms and herbs are employed for protection of the living.[5] The Evil Eye, or *mal ojo*,

4 Jonathan Roper, *Charms, Charmers, and Charming*, p. 180.

5 I have discussed these ideas at length in my essay 'Botanical Charms for Binding the Dead in European Folklore.' Though both nations have markedly different

nearly a universal phenomenon, is a prime example of this magical paradigm. The outpour of malefic emanation ascribed to the eyes of certain individuals, in some cases even without their knowledge, in known to transfix, infect, or drain the power from the victim beheld. Ancient Gnostic perspectives, particular those of Manichæism, hold particularly powerful examples of these concepts. Here, poison is acknowledged as a by-product of creation, but in encountering it, the work of the elect consists of continual purification, of 'liberating the light' which is co-mingled with darkness, and using the vessel of the body as well as the spirit to do so. Though bound by dualism, this view is important because of its classification of various kinds of darkness, poisons, impurities, or 'abortions.'

As concepts of purity and contamination affect the individual adept, so too they manifest in the collective body of initiates, where magicians are organized into a lodge or covine. Like the human body, a robust mystery cult possesses an active immune response. Part of this patterning arises from lore and customs of initiation, or of specialized means of recognition of appropriate potential initiates. Brethren may also serve as enfleshments of the old god Terminus, being the keepers of the boundaries. Inappropriate persons, encountering this immune system at its periphery, create specific astral interference patterns which facilitate their rejection. Likewise those appropriate to the work emanate certain harmonics which initiates are taught to recognize, such as the so-called 'Awen on the Brow' or 'Mark of Cain.' This magical immune function is analogous to the station of the Tyler in Freemasonry, the outer guard of the lodge whose charge is to prevent intrusion by the uninitiated, curious, or inappropriate.[6]

histories, their magico-religious subcultures are composite, given their shared history of global empire and hegemony.

6 Similar stations of appointment also exist in other esoteric orders.

Magical orders and sodalities, if they are living and vital, also possess their own 'karmas', amassed from the collective deeds of their initiates, in addition to the horde of atavism and ancestral force energizing the flesh of the present moment. Both the magical and mundane actions of each adept serve to advance the animated essence of the current, and feed it, or elsewise, in the case of parasitic action, degrade it. The moment of induction provides the new initiate with a great many new powers, but also the collective inheritance of the order's astral legacy, for good or ill. This is why transmutation of these karmas is a paramount magical concern, and the responsibility of every initiate. Pendell concedes that an important part of the work of the Poison Path is in 'undoing the mess left by the bad shamans', whose weapon is the Lie.[7]

This raises the shade of so-called 'cultural contamination' in magical subculture, the idea that foreign influences behave as corrupting toxins to a closed group. Amid certain rural magical lodges in both Britain and North America, I have encountered varying attitudes about purity and contamination; most often these relate to a fantasies of 'pure' nationalist or regional magical types, much of which are naked xenophobia. In these situations one must seriously question prevailing attitudes of 'culturally pure' magic where practice and lore have been syncretic for many generations, and influences such as Solomonic conjuration, Christian folk charming, Afro-Caribbean, and Native American lore have been locally adopted. Similarly, where magical practice has made its home in urban centers, one encounters a kindred bigotry advancing the primacy of democracy or egalitarianism, whilst simultaneously condemning and attacking all those found to be ideologically abhorrent. In both cases, the poison relates *prima facie* neither to interior and exterior, but rather to unconscious action and unexamined fallacies in individual reasoning, projected upon others. The phenomenon is not unique

7 *Pharmako/Poeia*, p. 4.

to esoteric orders, and indeed its endless variants may be discovered as an infectious by-product of human organization.

The power to reject or embrace the external is not merely the prerogative of groups. It is a power daily exercised by all individuals, though frequently in an unconscious manner. For the adept of the Poison Path, actions of inclusion or exclusion are of critical import. All such actions must be exacted consciously, knowing full well one's rationales. Mystery-cults have long been accused of elitism, or of concealing criminal activity, but in my own experience, the rationales for secrecy and exclusion are considerably more practical. More often, they arise from concerns of privacy and preserving intimacy among initiates, as well as barring the door to fools and criminals. This is a rationale operant as law to the individual; where it is violated, a profanation of the Self occurs; likewise the violation may occur at the level of a group, which is an association of individuals. Obviously, as with Poison itself, the negotiation between an interior and exterior mode of engagement forms a critical aspect of the work, and is allied with the Arcanum of the Sphinx.

As the Point at the center of a confluence of poisons and powers, the Adept will perpetually be called to develop his or her powers of discernment and self-discipline. Poison and its effects are systemic, thus its action is interpenetrative, as are the emanations of the most efficacious operations of the Art Magical. This implies embracing a perpetual consciousness of the magical body which may, at will, embrace or reject power, corrupt or purify, and the concept of the Magical Virgin.

Magical operations recorded in remotest antiquity often prescribe a virgin as the adept's ritual assistant, often in the capacity of a seer or in divinatory operations such as cleidomancy.[8] European grimoire texts frequently specify virgins for

8 Augury by use of a suspended key, auspicious when certain planetary powers are dignified in the zodiacal sign of the Virgin.

mediumship;[9] the practice of using a young virgin boy as the assistant to magical operation permeates Arabian sorcery and is still present in the modern era.[10] This 'unblemished' state, also common in sacrificial rites of ancient religion, marks conscious divisions of *numen* between 'purity' and 'contamination', and ascribes to them specific magical properties. The state of the body being metaphorically 'Edenic' further implies a state of receptivity that, like the liaison of First Woman and Serpent, assists the adept of the Poison Path.

In magical applications Zeroth or Void-mind states, though paradoxical by their very embodiments as concepts, are often the goal of specific magical praxes. Usually falling within the boundaries of asceticism, they are more often divertive from obsession or distraction than actual emptiness. Such psychostases are achieved by ritual techniques collapsing conception and awareness to their rudimentary components. Likewise a void-mind state may serve to stratify perception such that its center is receptive, whilst its periphery is protective: the very definition of the Magic Circle. If successful, the resultant state provides the most neutral media possible for impression during operations of the Art Magical, the better to provide the prepared vessel for encountering that which lies beyond it. When considering this magically generated 'ground', analogies with the painter's art are most apt: one can scarcely express the overwhelming and alien ecstasy experienced by the artist upon first beholding the blank canvas, a similar state to the Virgin Dyad first beholding each other in mutual desire.

Upon the poison path, we may speak of many 'firsts' – the initial sting or bite of a poisonous animal; the first adverse reaction from a prescription drug; the maiden experience of anæsthesia; the first visual apparition borne of an 'hallucinogenic' substance. This inceptive consciousness may also be applied to Astral Poi-

9 *The Key of Solomon the King*, Book II, Chapter II.

10 W. E. Lane, *An Account of the Manners and Customs of Modern Egyptians*, p. 95.

son: heartbreak, betrayal, emotional violence, nightmare. Each situation may be regarded as 'initiatic' in a technical sense, for all consist of a beginning, possessing unique powers under the monadic herald of inception.[11] If viewed from the perspective of tutelage, each virginal experience incepts a new pattern of teaching and amplifies the experience by optimizing a state of previous un-knowing.

The value of this embodied station is manifold. In Sabbatic Witchcraft, marked by the presence of a highly-sexualized magic in some applications, there is the notion of 'The Sabbatic Virgin', a zone of magical experience and activity where-in each magical interaction with the Other is a ravishing anew.[12] Importantly, this station demands the virtues of distinction, discernment, humility and the willingness to admit fallibility. It is thus the conscious enfleshment of Void by the adept in preparation for countless betrothals to power, an embodied practice I have come to call 'cultivating the perpetual neophyte.' Even in a realm of mastery, claimed as one's own and executed in deed, nothing is static, nor does the true adept ever cease learning. Thus may the Master embrace the station of 0°=neophyte, proceeding 'as if' each situation is entirely new, or on a basis whereby he or she may be challenged, proven wrong, or re-made in the Forge of Initiation. This is a critical aspect of Crooked Path Sorcery, and brings to mind the cyclical reversal of roles, such as that of slave and master, essential to ancient festivals such as the Saturnalia.

And yet the parameters of the Virgin are ever-shifting, subject to a great many conceptions, most importantly Her own. Whilst she may not have consummated the sexual act with a fleshly companion, she is certainly aware of her desire to do so, and of the tides of attraction, arousal, and physical compulsion, all of which are potent reservoirs of power. The touch of her own

11 In the Tarot, a quaternian manifestation of inceptive powers occurs in the four Ace cards of the minor arcana.

12 *Azoëtia* and *Lux Haeresis*.

hand will be well known to her, and thus the initiating force of First Agapæ arises in the pre-figuration of previous accretion of sensual knowledge and fantasy; the difference in consummation is in the presence of the Other. This body of self-knowledge, a combination of limited experience and vast imagination, is in every wise as important as each connubium which follows. Whilst, clinically, the inception-point of coitus marks a technical and spiritual terminus of the individual, it nonetheless is part of an eroto-magical continuum for the Virgin-Adept. This addresses the notion of preconception as a 'contaminant' to 'pure' magical experience.

Where the study of magical poisons is concerned, the station of the Magical Virgin, beyond all restrictions of the Self, is in fact determined by such factors as temperament, devotion, and embodied presence. The ability to attach with ferocious passion, and detach as readily, is also paramount. Here, we advocate no prudish avoidance on the basis of fear or morality, but an holistic recognition that each encounter with the unfamiliar may be a source of great power to the adept. One tenders oneself as bride or bridegroom unto Experience, wholly devoted in love, respect, and desire—the emblem of Sacrifice, whose lifeblood feeds not only the ancient gods, but one's own incarnation.

THE CORPOREAL LABORATORY

THE ALCHEMISTS OF ASIA, EUROPE, AND THE NEAR EAST pursued direct mystical knowledge through replication of the Universe in the microcosm of the laboratory. Some of the most dangerous compounds known to humanity were present in their operations, whether due to toxicity, volatility, or progressive corrosive action upon the human organism. This chemical pilgrimage occurred in the sealed and external proving grounds of cucurbits and retorts, furnaces and baths, according to the skill and subtlety of the Arts of Hermes.

The philosophers' own bodies also served as proving grounds for their reagents and nostrums. For many, this exposure was unintentional, the flesh serving as a filter for the toxic stew of mercuric vapor, reductive acids, and metallic sublimates. For others, like Basil Valentine, direct imbibition of transmuted metallic poisons, such as Antimony, was a deliberate and spiritual act.[1] The intersection of magic, mysticism, natural science, metallurgy, botany, and medicine collectively embodied in Alchemy implied a melothesic construct of the divine mysteries, the Corporeal Laboratory. By this route of inquiry, the powers of heaven and hell were embodied in the human corpus, as were the potentials of the earthly body as furnace, retort, *crucibulum*, and all the transmutative apparatus of the Philosopher.

This noetic path of corporeal self-discovery in relation to poison is exemplified by Dale Pendell in his trilogy of books *Pharmako/Pœia, Pharmako/Dynamis*, and *Pharmako/Gnosis*. By use of a kind of poetic empiricism, the Ally of each poison under investigation was made manifest to the adept by the act of its imbibition. This marked a return of needful subjectivity to the Poison Path, and the restoration of the Spirit of Discovery, an estranged and essential part of 'objectivity.' Inquiry was not limited merely to the routes of history, therapeutics, and chemical assay, but also through mantic embodiment of the poison genii themselves and their alignments with the adept's own magical trajectories. As an ever-present shade haunting, and at times directing, the exposition, the oracles were received from the poison *daimones* direct and revealed their important nature as trickster. As a critical theme of the works, Pendell rightly situates the mystical study of poisons within Alchemy, not only for its controlled work with deadly compounds, but also due to its heterodox occult pedigree and its rich tradition of poetry.

Prior to the flowering of alchemy, many of the compounds subjected to laboratory trials had a long history of human use. The substances present in most ancient cosmetics of the Near East reveal a number of beautifying substances composed of poisonous metallic oxides of mercury and lead.[2] This provides

1 *The Triumphal Chariot of Antimony* (1606), in which Valentine describes numerous prepared compounds of antimony, some taken orally.

2 For example, the classic Renaissance skin whitener Venetian Ceruse, also known

a link with the luminary Azazel, the Goat-headed One who in some elder accounts was the captain of the Watchers or Fallen Angels who rebelled against God, and led humankind astray with magic and forbidden arts. Each angelic reprobate governed a different dominion of magic, divination or technology, which he then taught to his human wife. Azazel instructed in both cosmetics and metallurgy, especially forging the weapons of war. The link between smithcraft and cosmetics is in the first instance alchemical-metallurgic: the smelting process promotes the natural emergence of oxide pigments prized for beautification. We may also view cosmetics and weapons of war as traditional armaments of each gender; poets have often described Feminine Beauty as a poison by which one is 'smitten.' Though there is significant theological and historical debate about the origins of the Watchers legends, a convincing case has been made that many of their features were borrowed from Greek legends of Prometheus, another transgressor of divine order and also figure aligned with the Alchemical art.[3]

The legends of transgressive gods stealing heavenly fire imply an important principle upon the poison path: knowledge must be embodied in order to become active and realize its potential. This embodiment is both passive (as with its situation in the mortal flesh) and active (as a continual wellspring of teaching from direct experience). In the legend of the Watchers, embodiment occurs by two important routes: through the 'earthing' of angelic beings after being cast out of heaven, and the angels' transmission of forbidden knowledge through direct teaching of human women. That these ancient liaisons of heaven and earth contain a strongly sexual component further underscores the theme of the reception and attainment of knowledge through

as Spirits of Saturn, derived from the toxic mineral cerussite or Lead carbonate, $PbCO_3$. Until relatively recently, white exterior paints contained a mixture of 75% lead carbonate and 25% oil.

3 George W. E. Nickelsburg, *Enoch 1 Hermeneia*.

the vehicle of the body. This is known in some forms of modern witchcraft as 'The Alembic of the Wise', an arcanum teaching the transmutative aspect of the body, culminating in its realization of 'The New Flesh' a magical paradigm uniting Self and Other in a dynamic and gnostic perichoresis.

III. *Saturn as the ouroboros-wielder, from Vincenzo Cartari's* Imagini colla sposizione degli dei degli antichi, *1581.*

Where experiential knowledge of poison is concerned, the legacies of Alchemy are many, not merely restricted to the shrouded byways of occult tradition. The antimonial cup, an ancient remedy persisting until the early modern era, is a prototype par excellence of the poison chalice-type *pharmakon*, but also of the use of the human laboratory for the transmutation of poison. Fashioned of pure metallic antimony, the vessel was filled with wine and allowed to stand a day, whereupon the toxic virtues of the metal leached into the mild ethanolic solution, creating a powerful emetic. Known also as *pocula emetica* or *calices vomitorii*, there is evidence that such cups, passed down in families from one generation to the next, accreted a kind of power or ritual significance not unlike saints' relics and the magical regalia of witches, cunning folk and Freemasons. The Paracelsian legacy of alchemical *iatrochemistry* endured in nineteenth-century pharmacopoeias in such official remedies as syrups of lead, White Lead, and mercuric chloride. The medical establishment's relation to these poisonous alchemical legacies is cyclical, with therapeutic metallic salts falling in and out of favor over time.[4] Still, the medical efficacy of such preparations has arisen in part by humanity's ancient and ongoing rôle as a test subject in a crucible of metallic venom.

In Astrology and Alchemy, poisons and their malign radiations are governed by the planet Saturn, the cannibalistic god of time, agriculture, and also the metal lead (and, esoterically, antimony). This association with lead, whose symptoms of gradual poisoning (d) afflict multiple organ systems, including the brain and nerves, is consistent with the iconography of the deity, often portrayed as old, crippled, or swathed in bandages and plasters. The Doctrine of Signatures, known for its complex arrays of magical correspondences, generally assigns poisonous plants to the dominion of Saturn, as well as those inhabiting low, boggy, and contaminated soils. Such places are the dwellings of the rotten, the verminous, and the dense precipitates of sublunary miasma.

Astrological aspects with Saturn are always concentrations of force, and often presage plague and famine; in times past the god was often evoked in planetary magic with a mixture of respect, fear, and horror. This underscores the god's astrological

4 A common mercury compound for topical use, mercurochrome or 'monkey blood', was a common household medicament in the United States until banned in 1998. Prescription medical use of gold and silver salts is currently on the rise.

station as a transpersonal planet, whose emanations have wider effects upon humanity as a whole, and govern transcarnative powers of the individual, such as the Śramanic concept of karma. This transection of the flesh of the present moment with past and future embodiments is atavistic in nature, and links Saturn with his old Greek forbear Kronus, lord of time. Saturn's governance of temporal cycles also accretes related concepts of decay, mortality, and the horror of personal demise, all kindred to poison and its associated complex of fears. An ancient Nabatæan enchantment reads in part:

> *I ask of Thee in Thy name, the kept one, the hidden one, in thy unending knowledge to hear my prayer, and to save me from the evil of Saturn, the old chief, the great one, the bright one, the luminous one, and from his killing, mortifying, and torturing my soul, and to keep life in my body.*[5]

Giordano Bruno's descriptions of Saturnian power are thoroughly taxonomic. Among the great host of the god's attendants are Putrefaction, Stench, Meagreness, Squalor, Indecency, Infamy, Poverty, Hatefulness, Sorrow, Vomiting, Stuttering, Crabbiness, Morosity, Severity, Impatience, Invalidism, Foolishness, Rigidity, Wrinkles, Toothlessness, delirious Imagination and Horror.[6] Each of these characteristics is applicable in its own way to poison; indeed some read like precise debilitative symptoms of intoxication.

The alchemical tradition furnishes us with a unique occult wisdom of poisons, both in philosophy and semiotics. The *caput mortuum* or dead head, being a residuum of the alchemical Work, is especially apropos the study of magical venoms, not only for its status as the excreta of the Grand Operation, but also its iconography of the skull. As with both Nature and the Alchemical

5 Ibn Wahshiya, *Book on Poisons.*

6 *On the Composition of Images, Signs and Ideas*, 1591 (1991).

Art, all conscious magical operations produce ætheric waste and detritus, and the art of discernment between them is of paramount importance.[7] When subjected to the proper fires of Art, the skull may become the vessel of future work, the 'purified body' which receives vivifying force. The process bears certain features active in the Mysteries of Christ: migration of the Body of the Savior from Golgotha, Place of the Skull, to the vessel of the tomb, through the flames of Hell's Harrowing, to the final Light of the Resurrected Body.

In the most rustic of environments this principle is active, true to the ancient words of Hermes, in his *Tabula Smaragdina*. In my youth, I was privileged to meet and study directly with a number of masters of herbalism to further my own knowledge of plants. One such master, an Englishman, shared with me an anecdote about having learned part of his craft amongst the hereditary fairy-doctors descended from the Welsh *Meddygon Myddfai*. He relayed that his hardened skepticism of folk medicine was strongly challenged when he witnessed a complete cure of shingles using the topical application of urine from a pig fed on certain plants. Likewise, feces and excreta of various kinds often form the basis of folk medicines.[8]

In Sabbatic Witchcraft, the *caput mortuum* finds its corollary in the vulgar, uninitiated flesh of the Profane Man, symbolized by the corpse of Abel after his murder by Cain.[9] It is the accursed matter, that which is rejected after the purifying fires of the Forge of Initiation. The so-called 'separation of the corpse from

7 *Caput Mortuum* was also a purple pigment used in oil painting—often, ironically, for rendering the vestments of religious personages. For this reason it was also known as Cardinal Purple. Also associated with oil painting is the malady known as Painter's wrist-drop, a peripheral toxic neuritis also called lead palsy, caused by gradual ingestion of soluble salts of lead common in many pigments such as Flake White.

8 For example, sheep dung used for baldness, *Irish Country Cures*.

9 Andrew Chumbley, *The Dragon-Book of Essex*, 1997; Gavin Semple, *The Devil's Noctuary* (1996–99), Daniel Schulke, 'Cainite Gnosis and the Sabbatic Tradition', *The Cauldron*, February 2012, and *The Psalter of Cain* (2012).

the angel' acts as the mediating bridge allowing passage from the circle of unrefined matter into the Circle of Art. Unlike usual Alchemical operations, the Death's Head is afforded certain legitimate powers rather than merely being rejected from the Labor or passed into a highly refined or purified form.

As it contains the antithesis of bodily transmutation, it serves as a concentrated Principle of Opposition, and is assigned to the dominion of curses or 'the Dark Body.' Thus, the black skull, as it sometimes called, possesses a malefic quality serving both the operations of healing and harming. In its bound and aright form, it may manifest as destructive force or blight; in its bound averse form it is a healing salvific—the medicine cup fashioned from the skull of the profane corpse. This recognition of a spectrum of power inherent in filth arises from an underlying ethos, pervasive in some forms of Traditional Witchcraft, emphasizing knowledge of right usage, rather than rejection or acceptance on a purely moral basis. Here, we return to the idea of embodiment as a necessary engine of magic and transmutation. As a feature of Crooked Path sorcery, the initiate is perpetually refining poison into nectar, a process sometimes described by its adherents as 'self-overcoming.'

In the modern era, toxic environmental contamination may be regarded as a *caput mortuum* an otherwise detrital consequence of advanced materials technologies such as chemical synthetics, nuclear power, industrial agriculture, and munitions manufacture. Their common aspect of the contaminants being hidden, subterranean, secreted, and lying just below the perceptible surface also resonates with the classic alchemical description of the estate of Nigredo, whereby the Stone conceals itself in the foulest filth. In Opposition, especially from the view of the Poison Path, we also recognize a similarity to the magical-alchemical formula of VITRIOL: Visita Interiora Terræ Rectificando Invenies Occultum Lapidem.[10] Though the example of industrial waste is clearly the result of human activity, it is important to note that, geologically, Nature keeps her poisons 'locked away' in the Earth's crust, either deep below the surface or in the form of less toxic minerals which must be smelted to extract the poison. Nature, in her 'pre-alchemic' state, offers one order of poison, but the Royal Art, or Science its bastard, may magnify it many times beyond.

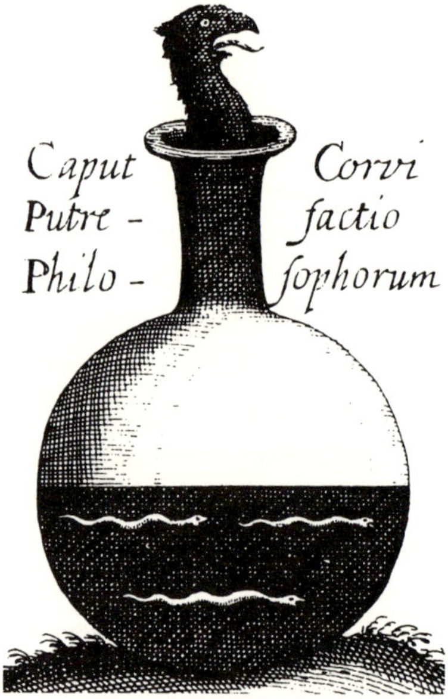

IV. *Alchemical Putrefaction, showing carrion bird, black matter, and worms. From* Anatomia Auri *by Johann Daniel Mylius, 1628.*

Industrial manufacturing paradigms have often rejected the principle that both product and by-product are dual emanations of a single operation, and that each bears its respective powers and responsibilities. From an alchemical perspective, such have given rise to a shadow-child of the Royal Art, a demonic alchemy which not only generated gold from the wonders conjured in its laboratories, but also manifold poisons which spawned a legacy of death, disease, mutation and sterility. Like the *qlippoth* or shadow-emanations of the Kabbalistic sephira, the husk-like impurities of industry do not merely cease activity once buried in the ground, but continue to leach corrupt influences over time

10 "Visit the innermost zones of the Earth and by Rectification shall you discover the Hidden Stone."

and space. Depending on the nature of the poison, their venomous nature may endure decades, centuries, or millennia and fuse with the strata of Nature herself. Persistent residue of the agricultural pesticide lead arsenate ($PbHAsO_4$) is a classic example. Once used extensively in orcharding, it persists in the soil for decades and can also be systemic, migrating into the tissues of trees and their fruit—a corrupt yet biologically healthy agricultural product aligned with the witches' poisoned apple bearing the mark of the fallen angel Samael.[11]

The modern era of recycling, in which an ethos of responsible re-use has been advanced in some areas, has provided a needful first step in reversing the damage. Yet those involved in the recycling industry are forced to admit that the process on an industrial scale generates pollutants of its own. This necessitates a viewpoint wherein a necessary part of fabrication is envisioning an object's architecture as both its alchemical *lapis* (the desired object) and its *caput mortuum* (its waste-product) before it congeals as a manifest form. Once thus aligned with the will of its creator, its future destiny is accordingly secured. Again, this dual ethos returns us to the Formula of Opposition, and the Crooked Path. Removing these principles to the Corporeal Laboratory, the entirety of the substance of the Initiate, fair and foul, must truly be tempered by its opposite, that a new and more powerful refinement may occur.

For the practitioner, all forms of poison, be they chemical, astral, emotional, or otherwise, provide opportunities for hermeneutic tutelage. Medical anesthesia, far from being an exact and safe procedure, is a prime example of this. With the sensorium catapulted downward into leaden twilight, both the body and its extension of consciousness trace the boundaries of Death's own circle; a small percentage of patients do not emerge alive.

11 The ability of Nature to incorporate deadly poison into thriving biological tissue over generations is remarkable, as is evidenced by the thriving marine life populations in Bikini atoll and other former nuclear test sites.

More importantly, the majority who do return from the ethers of anæsthesia frequently report visits to fantastic realms and intercourse with the dead, in many ways identical to well-known documentation of near-death experiences.

A surgeon associate of mine discovered, through a routine checkup, that he required a surgical intervention himself. In accordance with the privileges of his practice and his thorough knowledge of drugs, he was permitted to prescribe his own regimen of anesthesia for the operation, which included liberal levels of ketamine[12] in addition to other drugs. He later reported that the resulting waking delirium was suffused with otherworldly music, singing, and a seductive retinue of dancing nymphs which cavorted about him in the operating room. The features of this tableau superficially resemble some features of the classic iconography of the witches' Sabbat, particularly the so-called 'fairy Sabbaths.'[13]Similarly, my own anæsthetic experiences with clinically-regulated scopolamine have produced monstrous distortions of the sensorium which resembled the affected visual states of magical atavistic trance. This is characterized by a grotesquerie of proportion, and strange amalgams of beasts, men, women, and natural landforms.

An important 'spiritual legacy' of Alchemy, and of especial import to the Corporeal Laboratory, is the modern menagerie of synthetic psychoactives such as 2C-B.[14] Often dismissed by drug-takers as inferior to 'natural' drugs, or having no spiritual component, these novel substances properly belong to the genii of minerals, allied in spirit to the 'metallic consciousness' of Manuel DeLanda's philosophy of matter.[15] This particular spirit

12 Ketamine, $C_{13}H_{16}ClNO$, important as a veterinary and pediatric anesthetic, also as a recreational drug for its inebriant and hallucinogenic effects. Both therapeutic and non-medical usages have multiple applications.

13 Gustav Henningsen, 'The Ladies From the Outside.'

14 Here I speak of the Royal Art's virtues of Love of Sophia (Wisdom), exploration, innovation, and reverence in approach to Nature and God.

15 DeLanda, *Uniformity and Variability.*

of discovery, combining the purified marrow of empirical inquiry and direct visionary experience is exemplified in our current time by the chemist Alexander Shulgin (1925–2014). The originator of over 230 psychoactive and euphoriant compounds, he has extrapolated the reverent innovation of the alchemical philosopher into the chemist's laboratory, and wed it to the pilgrimage of Self. Each substance was thoroughly tested on himself and volunteer participants before a chronicle of its manifestation was published.

Applications of the Corporeal Laboratory have long been known to the witch-cult, although the name such magical formulae may take varies with time and place. The power which arises from its furnaces and bellows is chiefly atavistic and draws upon the great reservoir of the pre-incarnate body. Within, the most ancient wellsprings of fleshly power may be tapped by venerative and magical means, and empower the body into a proliferation of forms, each capable of manifold powers and hypostases. This is the so-called 'New Flesh' or Transmutation of the Body from a gross state to the angelic, wherein all possibilities are manifest through corporeal form, and the latent and vestigial attain radiance. This has historically been expressed in a variety of symbols, namely the beast-masked revelers of the midnight conventicle, and the theriocephalic forms, half human, half beast emergent as spirits by the light of the Circle's fire. The formulae may take on a multitude of extrapolations according to the desire of the witch, but a recurring pattern commands originative, formative and expressive phases. That is, the conjuration of raw atavistic power from the repositories of the flesh, giving it form, and breathing life into it anew even as it is presently embodied. Such techniques are aided by a variety of ecstatic sorcerous pathways, including the mesmeric, nostalgic, hypno-sexual, and routes of the Poison Path.

These formulæ may be accomplished by a single practitioner, but the force of the power is multiplied many times over in the Sabbat assembly. This is demonstrated by a magical operation known as Circulatum Sabbati, which magnifies and multiplies the mesmeric sexual power of eight adepts, comprised of four couplings of male and female, into a spirit-concrescence which is at once eucharistic and egregoric. This *octriga*, or coven of eight, is represented as a symbolic circle of eight conjoined alchemical retorts. The Elixir is generated in each individual retort, whose function is to cook and distill, and in doing so, the elixir passes from one body to the next in a repeated cycle, female to male to female, here seen as a conjunctive axis of solar and lunar. The force applied to each retort must be of concordant resonance, but the participation of multiple vessels, combined with perpetual cycling in emulation of the alchemical Circulatio leads to a processed fluidic medium which is many times more potent than a simple and separate combination of four 'alchemical weddings' of Sun and Moon.

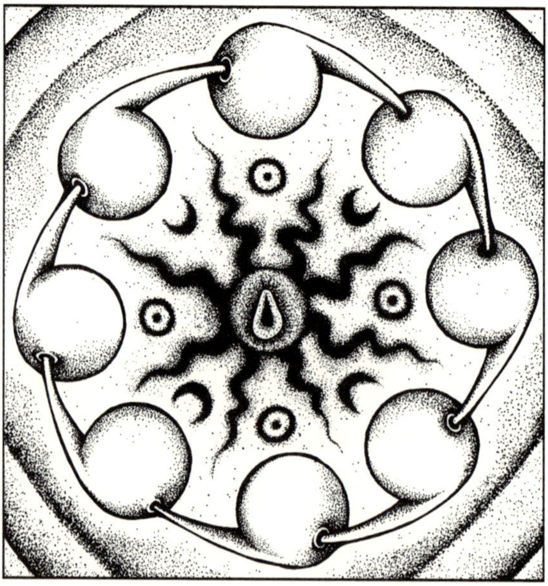

V. Circulatum Sabbati, *showing serial distillation of the magico-sexual powers of eight witch-adepts, as represented by alchemical retorts. Schulke, from* Ars Philtron *series of diagrams.*

The active, operant aspect of the Work demonstrates, through the Laboratory, that drinking the poison cup is easy; turning its black venom to gold is not. One of the highest and most essential responsibilities of the Poison Path is transmutation. Such a mandate necessitates a rubric whereby progress can be rightly judged and the mere appearance of transmutation separated from the Stone. Here one must heed the sages of old when they chose such words as 'skillful' and 'subtle' to frame the successful approaches to the Art. As Basil Valentine observed:

> *The bee extracts honey from the flower by the art which God has given to her; but when once the honey is visibly perfect, that sweet and fragrant liquid can be prepared in such a way as to become a potent and deadly poison. This is a fact which no one will believe who has not seen experimental proof of it.*

Despite the wisdom attained in experience, the use of the flesh for executing the operation of Art also presents vulnerabilities associated with subjectivity, namely mis-perception, and the pitfalls of addiction, tolerance, and over-intoxication. That which empowers and transforms the path of the adept must ultimately be held in greater esteem than that which concentrates existing constructs of illusion, delusion and dysfunction. More particular to poison, that Operation which effects a change in quintessence such that its power is no longer destructive to the vessel but creative is worthy of advancement to the next stage in the Work. Restriction ceases, motion ensues, and the dynamism of flesh and spirit expands in a reverberatory manner. The question must be asked: has the poison served as Muse or as Master? Has the Laboratory and its vessels become stronger, or more degraded? By answering such questions truthfully we may discern the Black Stone from the Green, and, perchance, witness the resurrection of the Dead Head into a living oracle of prophecy.

LEAVES OF HEKAT

HEKATE WAS THE ANCIENT GODDESS OF THE MOON, sorcery and of the night-traveled pathways, later syncretized with Artemis. The Greeks accounted her the daughter of Perseus and Asteria, and the mother of Scylla, the water-demon. Her center of worship was in Ægina, where annual mystery rites established by Orpheus occurred. Artistic representations show her bearing torches; where roads or paths met, a triple effigy of Hekate arose from masks placed at the junction, and she was venerated by a largely female lunar cult. Patron of the enchantress Medea, it was said that great honor came to those whose prayers Hecate received favorably, for she was regarded as a bestower of wealth. Her known period of worship was 800 BCE until the Christian era, but according to Gordon Simoons, the researcher whose book *Plants of Life, Plants of Death* addressed the subject of plant-taboos, her veneration continued much longer:

> *It is likely no accident that Hekate, chthonic goddess of sorcery who brought on or cured illness, was offered garlic in the form of a wreath to accompany the suppers provided her at cross-roads, which, as we have seen, were associated with her, and*

> *that Hecate was believed to punish with madness anyone who dared eat her suppers. Despite the rise of Christianity, Hecate and crossroads offerings did not disappear. Crossroads offerings persisted as late as the eleventh century, when there are reports of the Church attempting to put an end to them. Hecate herself, moreover, led the well-known witch ride of medieval times.*[1]

Persons who died early or unnatural deaths, such as suicides and murder victims, were "restless spirits flying abroad in the wind with Hekate."[2] As some archaeological yields have revealed, Hecate was invoked in conjunction with the use of humanoid fetishes known as *kolossoi*, sometimes for cursing; her rites including protective and exploitative magic, were practiced at crossroads. From antiquity to medieval times, Artemis-Hecate was strongly identified with the hordes of the dead, who traveled in procession during the Ember Nights, the twelve days from Christmas to Epiphany, and who were visible to certain seers.[3] Greeks called Hekate *triceps* or *triformis*; and she was thought to have three heads: that of a lion, a dog, and a mare; along with three goddessforms—Phœbe (the Moon) in heaven, Diana on Earth, and Proserpine in hell.

In Ovid's *Metamorphoses*, the sorceress Medea invokes "Thou, three-faced Hecate, who comest conscious of my design, and ye charms and the arts of the enchanters, and thou, too, Earth, that does furnish the enchanters with powerful herbs", the Thessalian sorceress Erichtho invokes Hecate in Lucan's *Pharsalia*. As we shall see, the invocation of Hekate emerges as a common feature of the magical arts of the Thessalian witches, as do a specific retinue of poisonous herbs.

1 Frederick J. Simoons, *Plants of Life, Plants of Death*, p. 143.

2 Hans Peter Duerr, *Dreamtime*, p. 193.

3 ibid., p. 35.

Though this magical reference occurs within mythical narrative, Hekate was invoked by folk magicians and sorcerers, and this is substantiated by the archaeological record. For example, PGM LXX4-25, a surviving Demotic Greek spell, invokes Hecate-Erishkagal for protection against daimones of punishment in the underworld. In addition to a protective incantation, the supplicant is instructed to simultaneously scatter sesame seeds about at a crossroad.[4] The 12th century scholar and religious reformer Eusthasisus of Thessalonika wrote:

> *...the name Empousa [is given] to the demon that appears to the unfortunate [at Hecate's bidding] under various forms or apparitions around midday, when solemn sacrifice is offered to the spirits of deceased relatives.*[5]

Her attributes as patroness of sorcery, providing protection, and having provenance over the dead and evil spirits, permeate the core of her deific power. They are also the fount from which the image of the Thessalian witch flows, the source of her power, and, arguably, later witch archetypes of medieval Europe.

Thessaly (or modern Thessalia) is the geographical region that occupies the central section of mainland Greece. Geographically isolated except for its eastern shoreline on the Aegean, it is surrounded by high mountain ranges encircling a low plain, and has a higher percentage of flatland than any other district in Greece. It borders Macedonia to the north, Sterea Ellada to the south, and Epirus to the west. From the earliest Greek writings, Thessaly has been regarded as a hive of *maleficia*, and famous Thessalian sorceresses from antiquity include such personages as Circe, Medea, Erichtho, Calypso, Dido and Folia. Their treatments in ancient histories are afforded the kind of simultaneous heroic respect and revulsion that leads one to suspect that

4 Betz, *The Greek Magical Papyri in Translation*, pp. 297–8.

5 Johannes Weyer, *Witches, Devils, and Doctors in the Renaissance*, p. 71.

the legends were grounded firmly in reality. In his *De præstigiis dæmonum,* a 1583 rational-skeptic treatment of the witchcraft phenomenon, Johannes Weyer equates the sorcery of the region with veneficium. He seems to sum up the old legends of Thessaly by remarking:

> *There is an ancient phrase, cited by Suidas, for the poisonings wrought by women: θεσσαλιχη γυνη, Thessalian woman, which we use, of course, because the Thessalian race, and especially the women, are notorious above all others on this account.*[6]

This ancient notion persisted from antiquity into the modern era. Paul Christian, author of the 1870 occult compendium *Historie de la Magie, du Monde Surnaturel et de la Fatalité à travers les Temps et les Peuples,* wrote:

> *The origin of witchcraft was very ancient; it began in Thessaly, a country as celebrated for its witches and wizards as for its Vale of Tempe or its Thermopylæ. The Thessalian women, who dedicated themselves to the profession of this spurious magic possessed, says Plato, the power to stop the sun in its course, to bring upon the earth the spirits of evil, to raise or calm tempests, to bring the dead back to life or kill the living.*[7]

Anacharsis, in his travels, observed these women working their magic, and noted their use of spells to kill bees and flocks of domestic animals, as well as the children of newly-married couples; particularly interesting was their use of wax fetishes pierced

6 Ibid, 560. Weyer goes on to qualify poisoning as being chiefly within the domain of women. There is one mention of male Thessalian sorcerers in Juvenal; writing in his *Satiræ,* he relays "this man produces magical charms, this other sells Thessalian philtres with which they [wives] can torment the minds of their husbands."

7 Christian, pp. 203–204.

with needles to exterminate people. Public superstition was so great that people who saw themselves modeled in this way believed they were destined to die, and often this terror shortened their lives.[8] In his *Elegies*, the poet Tibullus remarked of a certain Thessalian enchantress:

> *Now she makes the infernal hosts swarm round her with her magic screamings, and now she bids them be gone, sprinkling them with milk. When she pleases, she sweeps away the clouds from the sombre sky; when she pleases, she calls down the snow in summer by a word from her mouth. She is said to possess alone all the evil herbs known to Medea, alone to have brought the fierce dogs of Hekate under subjection. This witch has composed for me chants by which you may deceive all eyes.*

The powers wielded by these sorceresses were diverse. We are told that Medea blasted crops and fruit, and dried up springs; Horace tells of the enchantress Folia "who with Thessalian voice enchants the stars and moon and snatches them from the sky." Homer reports that Circe, by use of a golden rod and poisons, turned her spouse Picus into a colored bird. This transmutation of man into beast by means of sorcery was also accomplished by Circe on a grand scale with Ulysses' men. And Lucan, in *Pharsalia* 6, line 451: 'Through the charms of the Thessalian witches a love not induced by the Fates has entered into hardened hearts; and stern old men have burned with illicit flames.' As Anacharsis reported, these enchantresses had the powers to 'bring upon the earth the spirits of evil, to raise or calm tempests, to bring the dead back to life or kill the living,' a magical act firmly within the chthonic domains of Hekate.

Another reference of interest comes again from Anacharsis, who, having hidden himself nearby, observed their sorcery first

8 ibid., p. 204. Anarcharsis also noted of these witches that 'their sordid poverty equaled their ignorance.'

hand. The sorceress was spell-casting to retrieve the youth Polyclete, who had spurned his lover Salamis. The place of working was bedecked in

> *laurel-boughs and aromatic plants; some brass plates engraved with unknown letters; tufts of lambs' wool dyed purple; nails from a gibbet, still bearing traces of human flesh; skulls half-eaten by wild beasts; fragments of fingers, noses and ears, torn from corpses; entrails of victims stolen from temples; flasks in which was stored the blood of men who had died a violent death; a waxen figure of the goddess Hecate, painted white, black, and red and holding a whip, a lamp and sword intertwined by a serpent; several vases, some full of water from sacred fountains, others of milk and mountain honey...*

This squalid description may be exaggerated, but some of its components are familiar to the cult of Hekate, and, in modern times, to the practices of traditional folk magic charming and witchcraft. Porphyry wrote concerning how images should be consecrated to Hekate; they were to be surrounded with wormwood, and that domestic mice were to be painted, and "the finest ornaments such as were most pleasing to her, and so many mice as her forms were to be taken; then blood, myrrh, storax, and other things were to be burnt: which things if they were done, she would appear, and answer the worker thereof by dreams."

Another characteristic feature of the magic of the Thessalian witches was enchantment which relied heavily on the use of herbs, especially those considered to have a wretched, toxic, or baneful nature. In Ovid's The *Heroides,* Jason's lover Hypsipyle says of her rival, the sorceress Medea "By her incantations has she influenced thee; and with her enchanted sickle does she reap the dreadful plants." Virgil relates Circe changing Ulysses' men into pigs and asses by means of her 'potent herbs.' Ovid writes, in *The Making Up of the Face,* that "love touch us, and not through the strong herbs cut with the dreadful art of the witch's hand.

Trust not the grasses and mixed juices; try not the noxious poison of the lusting mare." The reference here is to hippomanes (ιπποµαεζ) or 'horse-rage'; perhaps the most famous ingredient in aphrodisiac potions among the Greeks of antiquity. It referred both to "the sluggish poison that flows from the pudenda of mares in heat" and to a small piece of flesh the size of a dried fig that clings to the brow of a newborn foal. It was usually incorporated into food or wine. In *The Golden Ass,* Apuleius described the sorceress Pamphile in this way:

> *She is a magician of note; and she is believed to be mistress of every sepulchral enchantment. By breathing upon sprigs and pebbles and other trifling objects, she knows how to plunge the entire light of this starry world into the depths of Tartarus and into the Chaos of old.*

Also curious was the mode of gathering these herbs. Many of the sorceresses gathered their plants by moonlight. The full moon, according to Weyer, infects the herbs gathered "with its evil froth." This recalls Phebe, Hekate's lunar aspect. Virgil also remarked that "flourishing herbs are sought, clipped by moonlight with brazen sickles,"* Weyer argues that this was because it was believed that bronze possessed medicinal powers, and therefore the witch cut the herbs with brazen sickles rather than iron. There are also widespread European beliefs that iron is annoying or poisonous to evil spirits; thus it is possible a sorceress working with such spirits as familiars would necessarily degrade her magic by use of iron.

Herbs used by these practitioners of the magics of Hekate were sometimes collectively referred to as the "Pontic herbs." Virgil, in his *Eclogues,* remarked:

> *Mœris herself gave me these herbs and these poisons culled in Pontus; many such things grow in Pontus. I have often seen Moeris become a wolf by their means and hide in the forests,*

> *and rouse souls from the depths of the tomb, and move the planted crops from one place to another.*

Here we not only find a reference to herbs and poisons in the same source, but also (in addition to translocation of crops) two types of magic they give rise to: transformation from human to animal, and the summoning of the dead. These, in addition to Erichtho's love-charm as mentioned by Lucan, are of interest to us in examining what specific herbs the sorceresses of Thessaly may have been using. Pontus was known to harbor many species of toxic plants, some of which resulted in mass-poisonings. An example of this occurred in 400 BCE, when a Greek mercenary army returning via Pontus was poisoned by Mainomenon, a honey whose chief source of pollen and nectar was a toxic Rhododendron species. A large number of the soldiers fell unconscious, and some perished. Two of the poisonous species endemic to the area are *Rhododendron luteum* and *R. ponticum*.

In his examination of the magico-shamanic consciousness of medieval Europe, German scholar H. P. Duerr remarks that "There is the tradition that Circe mixed some sort of 'bewitching juices' into the *kykeon*, which was prepared from grated cheese, flour, honey, and pramnic wine, so that Ulysses' men forgot their homeland and changed into pigs...In the later literature we read on occasion that poppy, mint, rue, and coriander were mixed into the kykeon..."[9] Kykeon is ancient Greek for 'mixture' and was also used for the ceremonial brew administered at the mystery Rites of Eleusis.

One favored plant often associated with Circe's brew in the research of magical poisons research is henbane (*Hyoscyamus* spp.). A hairy, sticky, odorous biennial or annual plant, it is found in wastelands, rural byways, and sandy areas of Europe and also naturalized in North America. It is rich in tropane alkaloids, including atropine, hyoscine, hyoscyamine, and scopolamine,

9 Duerr, p. 208.

a group of compounds important to European witchcraft and which entheogenic plant researcher Jonathan Ott has collectively named the 'visionary tropanes.'[10] Taken internally, henbane specifically effects the central nervous system: initially stimulation, soon followed by confusion, stupor, and disturbing visions. This is almost always accompanied by the classic side-effects of the Nightshades: dilation of pupils, a gradual drying of all bodily humors, reduced peristalsis, and vasodilation accompanied by increased blood pressure. Overdose gives rise to central narcotic paralysis, coma, and death.

The ancient Greek name *Hyoskyamos*, from which the plant's botanical nomenclature derives, means 'hog-bean'; it has long been conjectured that the name refers to the Circe legend of magically transforming men into hogs. However, as the Danish plant historian Harold Hansen notes, the German herbalist Otho Brunfels (1488–1534) claimed the name was a reference to the fact that pigs get cramps when they consume henbane. The common name references the fact that birds foraging on the wild seed quickly become stupefied, and are more easily hunted and killed, a phenomenon observed by modern hunters in areas where the plant grows wild.[11] Despite the debate about the origins of the name, Hansen's position with regard to henbane as used by the Thessalian witches is clear:

> *In the works of Apollonius Rhodius...as well as of Homer, there are stories of magic drinks which indicate that the alkaloid hyoscyamine was the most active ingredient. The first two authors mentioned above give detailed descriptions of the way in which Circe and her brother's daughter Medea, who was equally skilled in the magic arts, gathered their magical herbs. There can scarcely be any doubt that these to sinister mythi-*

10 Ott, *Pharmacotheon*.

11 The author thanks Cody Dickerson for permission of reference.

> *cal figures were based on actual people—contemporary Greek witches who knew the dangerous powers of henbane.*[12]

Hansen argues that the two species employed in ancient Greece were golden henbane (*Hyoscyamus aureus*), and white henbane (*H. albus*), as they were localized there; this does not rule out the better-known black henbane (*H. niger*), or Egyptian henbane (*H. muticus*). The latter is recognized as the more powerful species, and is grown commercially as a source for pharmaceutical hyoscyamine. Dioscorides was clear that *H. albus* was more desirable as a therapeutic medicine because it was less likely to cause madness; perhaps making it simultaneously less desirable as an herb for bewitching others.

In terms of the applicability of henbane's pharmacology for sorcery, it is indeed more suited to shape-changing spells involving the magical transmogrification of man into beast. Grieve cites the case of a monastery in which the monks mistook the roots of henbane for chicory wherein the monks were "attacked with a sort of delirious frenzy, accompanied in many cases by such hallucinations that the establishment resembled a lunatic asylum." Hyoscyamine, along with the tropane alkaloid scopolamine, have long been known to inspire visions with strong phantasmagoric or erotic content; delirium, frenzy, and varying degrees of amnesia upon return to the senses. A poem by the 5th century Roman philosopher Boethius recounts the shape-changing sorcery of Circe:

> *With a hand skilled in the use of herbs, she changes these guests in various ways. One of them has assumed the countenance of a boar. Another, as a Marmaric lion, grows long teeth and claws. This other, recently joined to the race of wolves, makes ready to weep but howls instead; and yet another, as an Indian tiger, walks tamely though the house...These poisons more effectively*

12 Hansen, *The Witch's Garden*, pp. 44–45.

> *lead a man away from his true self—the dread poisons which enter deep within and harm not the body but savagely wound the mind.*

Such descriptions are consistent with other accounts of poisoning by nightshades containing scopolamine and hyoscyamine, including my own experiments with henbane, wherein a feral, 'hyperkinesthetic consciousness' arises, a kind of muscular cognition which functions as a sensory organ, specifically with relation to perception of movement at one's visual peripheries and imminent threat.[13] Additionally, henbane is also known historically as a frequent addition to love philtres.

The 'visionary tropane' alkaloid complex occurs in many European solanaceous plants, importantly the mandrake (*Mandragora* spp.), belladonna (*Atropa* spp.), and *Scopolia* spp., and these, as well as other nightshades, should not be ruled out when considering possibilities for the Thessalian pharmakopœia.

According to Pliny, Thessalian witches employed black nightshade (*Solanum nigrum*) in a magic philter to evoke "obscene desires, forms, and images." Mandrake bore the folk name *Circieum* in some regions. Challenging the supposition of R. de Ropp that the Dinoysian maenads of antiquity drank wine laced with belladonna, Duerr notes that the plant was rare in ancient Greece, found only in a small number of woods in Thessaly; this, however, would not prevent the magicians of Thessaly from using it in their rites and philtres. The plant's reputation in sorcery is well-established; the priest-physicians of ancient Sumer employed it for driving out disease-causing noxious spirits; and in his 1589 *Magia Naturalis* Porta wrote of it:

13 Schulke, "Riding the Solanaceous Nightmare," *Esoterra: A Journal of Extreme Culture*, 1998. The article was an early and rough recension of material from *Hypnotikon*, my personal formulary of work in this theatre.

> *A drachm of the root of which, among other properties, has this, that it will make men mad without any hurt. So that it is a most pleasant spectacle to behold such mad whimsies and visions, which is also cured by sleep. But sometimes they refuse to eat. Nevertheless, we give this precaution, that all those roots or seeds which cause the takers of them to see delightful visions, if their dose be increased, will continue this alienation of mind for three days. But if quadrupled, it brings death. Wherefore we must proceed cautiously with them.*

Belladonna is easily cultivated, and even if Thessalian sorcerers did not grow it, tropane alkaloids survive the drying process where desiccated material may have been available. thorn-apple (*Datura stramonium*), while possessing the necessary virtues to effect theriomorphic and amatory sorcery, was almost certainly a latecomer to the European flora, thought to have been brought by the Romany from India; the earliest references in European herbals occur in the early to mid-sixteenth century.

Thessalian witches were credited with necromancy, that is, raising the dead. Of this type of magic, there are often classified two types: the first, known as *sciomancy*, is the raising of the disembodied spirit of the deceased, either to visible or auditory appearance or by mediumistic means; the second, *necyomancy* is the re-animation of dead tissue, such as those works of necromancy attributed to Jesus of Nazareth. Though Thessalian witches are documented as using both types of necromancy, the first is more common and the type we shall discuss in the context of plant-magic, involving aconite and poison hemlock.[14]

Greeks held poison hemlock and aconite sacred to Hecate, and that both plants have intimate associations with death. Diodorus reports in Book 5 of his *Ancient History* that Hekate first discovered aconite and "expended all her efforts in making up

14 Reanimation of a corpse is accomplished by Erichtho in Lucan's *Pharsalia* (6, lines 750–2).

deadly poisons." Aconite, also known as monkshood and wolfsbane, was believed by the ancient Greeks to have sprung from the banks of the Acheron, a river in Hades; and, alternatively, from the slavering jowls of the hell-hound Cerberus. Aconite is also supposed to have been the poison that formed the cup Medea prepared for Theseus. Plutarch gives an account of Marc Antony's army being in want of food, digging up some of these roots, and devouring them with the result that every man dying in paroxysm. In some myths, Hekate poisoned her father with aconite.

VI. *Yellow Monkshood* (Aconitum lycoctonum)

Correct botanical identification of the ancient Greek aconite is still unresolved. There are a number of species native to Europe besides *Aconitum napellus*, which is generally regarded as the most potent and certainly the best known. European species include yellow wolfsbane (*A. lycoctonum*), as well as *A. compactum* in the Western Alps; *A. orientale* is known in the Caucasus Mountains, and *A. paniculatum*, *A. tauricum*, *A. variegatum*, and *A. vulparia* are all common to southern or southeastern Europe. What most aconite species share in common, particularly in their non-magical applications, is a propensity for causing death if administered in the correct amounts. In the time of Theophrastus, it was a crime to own aconite plants, so feared was its venom and the poisoners who would employ it; this is an early example of drug prohibition. Chemically, aconite contains a range of diterpene alkaloids including mesaconitine, hypaconitine, non-nitrogenous compounds including salsolinol; picratonitine, aconine, benzoylamine, and neopelline. The roots and leaves contain the greatest concentration of poisons. These toxins possess a characterology of envenomation which clearly echoes not only the proclivities of Hekate, but of the dead and necromancy itself. The ancient Greek writer Nicander of Colophon (c. 270–200 BC) wrote of aconite in his *Alexipharmaca*:

> *When one takes aconite, the drinker's jaws and the roof of his mouth and his gums are constricted by the bitter draught as it wraps itself about the top of the chest, crushing the man with evil choking in the throws of heartburn. The top of the belly is gripped with pain—the swelling, open mouth of the lower stomach, which some call the 'heart' of the Digestive Vessel, others the 'receiver of the stomach—and the gate is closed immediately upon the beginning of the intestines where a man's food in all its abundance is carried in. And all the while, the moisture drips from his streaming eyes; and his belly sore shaken throws up wind, and much of it settles below about his mid-navel; and in his head is a grievous weight, and there ensues a rapid throbbing beneath his temples, and with his eyes he sees things double...*

Modern toxicologists agree on these symptoms, with one curious addition: those who survive aconite poisoning report hallucinations and bizarre sensory disturbances for some time after the poison has been excreted. It would seem that, even if we lack clear references on the use of aconite in Thessalian sorcery, the plant was known to the Greeks sufficiently to attribute it to their goddess of the dead and sorcery. Given what is known of the powers and idiosyncrasies of aconite poisoning, it possible, even probable, that this plant may have been used as an adjunct to rites calling forth the Mighty Dead.

Aconite has potential as a candidate for other types of ancient Thessalian sorcery, such as love-philters. In Taoist sex magic, Fu-tzu or Chinese aconite (*A. carmichaeli*) was combined with oyster shells, travertine, and other herbs to create the much esteemed Hanshi, or Five Mineral Powder, a stimulating aphrodisiac taken in brandy. It is a frequent characteristic of plants known for their powers of lethal poisoning that minute doses have different effects, such as the stimulating effects of strychnine from *Strychnos nux-vomica*. Aconite, along with poison hemlock and nightshade plants such as belladonna, is also frequently identified with medieval witches flying ointments.

A final consideration of the Thessalian pharmacopeia is poison hemlock (*Conium maculatum*), the presumed agent by which Socrates was forcefully ushered out of this world. The ancient Greek poison *Conium* was derived from a plant called *koneion*, whose precise botanical identity is still unknown to us. Dioscorides left a description of the plant, which is certainly a member of the Apiaceae or Dill Family, but Linnaeus, when creating his system of plant taxonomy, assigned the generic *Conium* to this plant. According to Plato, there were no violent symptoms of Socrates' demise, an observation at odds with what we know of hemlock poisoning; it is probable that opium was an adjunct to the suicide-potion for its narcotic and anaesthetic effects. Like aconite, poison hemlock was also considered sacred to Hekate by the ancient Greeks, and appears in a number of medieval recipes for 'witches' flying ointments.' Harold Hansen in *The Witch's Garden* writes:

> *Chemical experiments have proved that the poisons of the two hemlocks [*C. maculatum *and Cowbane,* Cicuta virosa*], especially coniine, taken in small doses and rubbed into the skin, can produce the sensation of gliding through the air.*

As we have already seen, Hekate has been associated with the flights throughout the air of the dead. This link with the dead, through the medium of drug-induced flight, or by employing Conium in conjunction with trance techniques, could indeed have been part of the sorcerous necromantic repertoire of the Thessalian witches, and their spiritual ancestors, the *lamiæ* of the Middle Ages.

THE

SPIRIT MEADOW

THE NAME *ELYSIUM*, THE PARADISIACAL SPIRIT-FIELD OF Greek lore wherein the shades of the blessed dead dwelt without strife, is thought by one etymological derivation to originate in a word meaning 'to be deeply stirred from joy.'[1] These ethereal meadows were described as possessing curious metaphysical properties, being both a source of bliss but also existing 'in no fixed place.' Indeed, standing alone in the vastness of tallgrass meadow, its plain stretching to the horizon, one experiences an intimation of infinity, and a state of mystical reverie incepted by vast absences. When stirred by the wind, the prairie grasses bend and sway in the air, creating wave-patterns evocative of the *daimones* or spirituous intelligences of the field. Immersed in this vast and pulsating locus, the phenomenon can be both hypnotic and ecstatic.

1 Henry George Liddell and Robert Scott. *A Greek-English Lexicon*.

The meadow, that most idyllic and pastoral of habitats, is generally an open field or clearing, whether in high alpine zones, steppe, or bottomland. Comprised largely of mixed grass species, pulses, and their botanical allies, meadows are rich in plant and animal biodiversity if occurring in their wild state. The meadow is the basis for the more expansive prairie or *pampas*, and also the wild habitat which ultimately gave rise to its well known domesticated counterparts: the agricultural field, pasture, and greensward.[2] In certain spiritual and occult traditions, particularly witchcraft, the meadow is also a locus rich in ecstatic power, and a zone of hidden poison and monstrosity. This chapter will examine the intersection of the place of the meadow with the heretical and complex weaving of poison, witchcraft, and *phytognosis*, or altered states of consciousness incepted by human use of psychoactive plants.

In addition to crossroads, caves, and forests, the meadow, as a general feature, is mentioned in medieval witchcraft lore as a place where the Witches' Sabbat itself takes place. According to the Spanish Inquisition, *Aquelarre* was both the name for the Sabbat of the Basque witches, and the place of their gathering. Inquisitors derived the word from *aker*, a male goat, and *larre*, a meadow, but it may have originally been *alkelarre, or* 'the meadow of *alka*', the local name for *Dactylis hispanica*, a poisonous cock's foot grass known in the area. Likewise, the notorious Swedish witches of Mora held their assembly at the Blåkulla, a place described by their inquisitors as 'a delicate large meadow, whereof you can see no end.'

In reading such descriptions of the place of the Sabbat, one recalls the meadow in association with the location of other

2 The lawn or domesticated meadow, though a more commonplace type than the wild meadow, presents a multitude of poisonous implications. Lawn pesticides are notoriously toxic and account for a great percentage of the agricultural poisons of commerce. Likewise, modern management of pasture and grazing lands have also given rise to their own ecological problems, such as the introduction of invasive grasses.

night-travelling sorcerers in European history, the best known of which are the *benandanti* of the Italian Friuli, who waged magical war for the fertility of the fields. The Ossetian *burkudzäutä* maintained that they fought on the 'meadows of the beyond' to "wrest the shoots of wheat from the souls of the dead."[3] The high meadows of the Allgäu in the Swiss Alps were a frequent gathering place of the otherworldly *Nachtvolk*, or People of the Night, who bore a resemblance to the faerie host and performed strange nocturnal rites.[4] As scholar Carlo Ginzburg has demonstrated, as well as others who have built on his research, the characteristics of such hidden societies did not easily conform to clerical stereotypes of the witches and their Sabbats, despite the fact that the official records conflated them.

In modern occult practice and philosophy, the meadow also plays a small but significant role in the practice of Anglo-American folk magic, in the tradition of Sabbatic Witchcraft. This syncretic current includes oneiric revelation and dream incubation as important parts of its ritual practice. Within this occult framework is a philosophical construct known as the 'Spirit Meadow', sometimes described as a four-sided field bounded by the states of sleeping, waking, dreaming, and trance; it is simultaneously considered a state of embodied power and an actual place. The Meadow is characterized in part by an effusion of the sensorium wherein the body experiences rapid flight over a meadow, or a field or corn. This sensation occurs as a refraction through one of more of the four magical zones which border the locus, followed by an 'indwelling' of the practitioner within the midst of the meadow itself. This 'flight', especially as experienced via the dreaming medium, may also include ærial transport on a vari-

3 Carlo Ginzburg, *Ecstasies: Deciphering the Witches' Sabbath*. (New York: Penguin Books, 1991, pp. 193–194. In the village of Soccia, the spirit-battles that ensued were fought with branches of Asphodel, "the plant which, according to the ancients, grew in the meadows of the netherworld."

4 Wolfgang Behringer, *Shaman of Oberstdorf: Chonrad Stoekhlin and the Phantoms of the Night*.

ety of mounts, including animals, objects, humans, trees, and supernatural entities. In this ontological framework the Spirit Meadow is considered a 'place that is not a place' that serves as both a state of orientation for the dreamer and also the fecundated locus that gives birth to dreams themselves.

In addition to its historical reference as Aquelarre, the meadow of the Sabbat has also been referenced as the Goat-Fold of Azazel, the Heath of the Billy-Goat[5] and the Vale of Josaphat. The zone is also linked in the customs of Sabbatic Witchcraft with the Field of Meimun, the legendary steppe where roamed the First Horse tamed by Cain. Known by many names in the Sabbatic Tradition, the ancestral shade of the First Horse is sometimes called *Eokharnast*; the act of its taming is a complex metaphor of the relationship between the sorcerer and divine power.[6] This is further linked with the concept of Cain as the outcast bearer of forbidden power and the protosarkian ancestor of witches.

Grain, being the product of the domesticated meadow, is also interwoven as a part of the magic of witchcraft. The magical destruction of farmed crops was a frequent witchcraft charge, and an act sometimes carefully depicted in early modern witchcraft iconography. During the period of the European witch trials, crop failure—mostly that afflicting cornfields, vineyards and orchards—frequently led to legal accusations of crop blighting. Folk magic, whether Christian or otherwise, often concerned itself with protection of the farm, be it crops or livestock, often from witchcraft or other devilry.[7]

5 Heathland, such as characterizes the Brocken, is a different ecological zone than the meadow, but equally implicated in historical accounts of witch-gatherings. According to DeLancre, the place of the Sabbat is called The Heath of the Billy Goat 'because the first places where these assemblies were discovered were moors or heaths, on account of the open space, and because many goats, sheep and other similar animals are seen there.' DeLancre, *On the Inconstancy of Witches*.

6 See *The Psalter of Cain*, Xoanon, 2012.

7 In the case of Sabbatic Witchcraft, the historical point of emergence of the Spirit Meadow likely relates less to medieval constructs of the Sabbat, and more to

Likewise, the malefic spells witches were accused of often involved the actual use of grain. One particular spell related in an anonymous inquisitorial tract of 1450, creates a powder milled from a dried catskin, stuffed with barley, rye, oats, grapes, and other plants. The powder was scattered on the wind to blight crops; not only were many of its critical components grains, the very intent of the spell was linked to attacks on the corn.[8]

There is also a curious relationship between witches, grain and the Devil which seems to have manifest as a kind of 'Devil's Tithe.' A woman tried in 1432 in Faido confessed under pressure that she had invoked the devil at noon on the bridge of Altanca, who appeared in the form of a goat and granted her a quantity of food. In return, she paid him a third of the new-mown hay lying in the fields of Val Levantina. Another instance was documented in an anonymous publication of 1460, where it was reported that witches were compelled to pay a monthly tithe of grain to the Devil, or else receive from him a harsh beating.[9] These offerings of fruits of the meadow to the Evil One are resonant with the Devil's Portion, a time honored practice of rural England. The custom demands that, when planting, a portion of the seed is offered to the Devil to ensure a good crop:

Four seeds in a hole
One for the birds
One for the mice
And one for the Master.[10]

the tradition's evolution in grain-growing areas of England, particularly Essex and neighboring counties. Certain rituals pre-dating the First World War, such as traversing the borders of fields while mounted on a horse-headed rod, testify to the importance of the field, and its intersection with the lore of the Sabbat.

8 *Errores Gazariorum*, c. 1450, quoted in *Lea*. Vol. 1 p. 274.

9 *La Vaudereye le Lyonois en brief.*

10 *Salt of the Earth: Origins and Meanings of Country Sayings*, p. 81. The Devil having dominion over grain also recalls the strange phenomenon of the 'Hertfordshire Mowing Devil' of 1678, as well as the alleged sorcerer George Pickingill

Variants of the charm are numerous, and reflect the farming realities that growing grain is far from a sure bet: disease, drought, foul weather, pests and other calamity could precipitate a disaster for the crop. The Devil's ransom of grain also recalls the traditional English protocol requiring a farm to surrender a tenth of its yield to the parson, who gathered it up in a specially constructed *tithe-barn*.[11]

In the traditions of Sabbatic Witchcraft, the figure of the *Devil-in-the-Grain* is found in the deific form of Mahazhael-Deval, an angelic-demonic instar of the Corn God which predominates in British agricultural lore. In this cycle of myth, repeated in various agrarian societies, the Fallow Field, Seed, Ripened Field, Harvest, and Loaf are personified as arcana of a singular entity, one which undergoes ritual slaughter for the sustenance of all, but is resurrected in a transmuted form.

In addition to its status as a location of the witches' Sabbat and otherworldly spirits, I shall also presently use the term 'Spirit-Meadow' to refer to a cabal of plants, linked by toxicity and their propensity for inducing delirium when ingested, which inhabit fields and meadows. These dwellers of the field have long possessed both sinister and sordid reputations, and are known not only for their disruption of agriculture, but also their effects of mental derangement, likened in some cases to demonianism and ecstatic frenzy. In sum, they exist as the Inverse Field, the shadow-grains which haunt the golden rows of John Barleycorn, and comprise a theoretical but highly specific witches' pharmacopœia.

The botanical designation Poaceae, or Grass Family, is a massive grouping of plants including much of the important cereal crops of the world. These include wheat, barley emmer, triticale, rye, and oats; grains which played a crucial role in the develop-

of Canewdon, Essex, rumored to have used his imps to cut a field of corn in only half an hour.

11 George Ewart Evans, *Ask the Fellows who Cut the Hay*, p. 98.

ment of human civilization, and, arguably, its physical and spiritual evolution. Central to the Mystery of the Bread, the grains are not only our companions in sustenance but also in mysticism: the sacramental loaf, often regarded as solely a Christian mystery, predates the cult of Jesus.

The Poaceæ also contains a large number of weeds—grasses which the noble agricultural art has disowned—and now exerts great strength to eradicate. Among these is Darnel (*Lolium temulentum*), a pestilent plant which over time has evolved to infect cultivated Wheat. Also known as Poison Ryegrass, Darnel mimics cereal crops in both form and growth habit, to the extent that

VII. *Darnel* (Lolium temulentum). *L. H. Pammel, from* A Manual of Poisonous Plants *(1911).*

one of its old common names is False Wheat.[12] Its perpetual infestation of cultivated fields has a long history in human agriculture: *L. temulentum* is thought to have co-evolved with wheat in the Fertile Crescent, and its seeds have been identified in the archaeological remains of many ancient granaries, some dating to c. 8,500 BCE. Industrialized agriculture has largely pushed Darnel to extinction, or to the peripheries of fields, but in developing countries it still thrives where grain crops are farmed, especially amongst those that are hand-sown.

Darnel toxicosis usually includes such symptoms as vertigo, confusion, disorientation, stupefaction, emesis, loss of motor control, fatigue, delirium, convulsions and coma. Many of these symptoms resonate with the frenzied and phantasmic effects of the Witches' Sabbat, as well as trance-possession in general. Historically, Darnel poisoning usually occurred by consuming bread or ale made with cereal grains infested with the plant's seeds. Flour fouled with Darnel seed was typically gray in color and possessed a pungent and unpalatable bitterness. According to Gerard's *Herball* of 1597:

> *The new bread wherein darnell is, eaten hot, causeth drunkenness; in like manner doth beere or ale wherein the seede is fallen, or put into the mault.*

Darnel may well have been present in the notorious 'black breads' served at the Sabbat, as documented by DeLancre,[13] either as a deliberate inclusion in the bread for its deliriant qualities, or simply as a by-product of the agricultural practices of the time. It is plausible that a toxic stew of deliriant grains, serving as the

12 Its mimesis in relation to favored grain crops corresponds with the shape-shifter tradition of witchcraft and sorcery.

13 According to DeLancre, "...their bread is some horrible black cake made of black millet and some other drug..." which 'confuses the senses' of those who eat it, and likewise binds them to Satan.

component basis of the Sabbat-bread, had synergetic effects operating between psychoactive components, if indeed darnel formed a part of witchcraft rites. This idea has been proposed in relation to the everyday bread eaten in medieval Europe, by the Italian scholar Piero Camporesi, who devoted an entire book to the subject.[14]

Much of our historical knowledge about darnel intoxication is derived from cases of accidental contamination. However, *L. temulentum* also has a long history of deliberate use as an intoxicant, similar to the use of clary sage (*Salvia sclarea*) for increasing the stupefying powers of beer and other alcoholic beverages. This was observed by W. T. Marchant as recently as 1888, when he reported that

> *Darnel, or lolium timulentum, which is vulgarly known under the name of sturdy when malted with barley, a process which the seeds of it often undergo, causes the ale brewed from it to be speedily intoxicating. It produces the same effect when mixed with bread and eaten hot.*[15]

Phytochemistry of darnel and its intoxicating effects on humans is not well understood by science and is subject to frequent revision. The stupefying and frenzying effects of *Lolium*, as well as much of its genetic adaptations, are due to the presence of an endophyte—a cohabiting organism within the grass itself—in this case an ascomycete fungus of the genus *Epichloë*, itself a member of the same family as Ergot (*Claviceps purpurea*). To the Greeks it was an inceptor of frenetic *ecstasis*, as implied by its name αίρα (*aira*) or 'plant of frenzy.'

Pearl F. Braude notes that Chaucer in *Canterbury Tales*, associates Lollard heretics with Lolium, the name for darnel; draw-

14 *The Bread of Dreams.*

15 *In Praise of Ale: Songs, Ballads, Epigrams and Anecdotes Relating to Beer, Malt and Hops.*

ing metaphorical linkage between a Christianity infested with heterodoxy and a grain-polluting agricultural pest.[16] This was further applied in medieval theology to Cain, seen as an ancient antecessor of heretics. Like heresy, its physical and philosophical designation as a weed places it beyond the margin of what is acceptable, in a state of exile. Indeed, in the biblical 'Parable of the Tares', this infesting role is associated with Satan:

> *The field is the world, and the good seed stands for the people of the kingdom. The weeds are the people of the evil one, and the enemy who sows them is the devil. The harvest is the end of the age, and the harvesters are angels. As the weeds are pulled up and burned in the fire, so it will be at the end of the age.*[17]

Taking the ancient Christian idea of heresy as an infesting or contaminating weed, the philosophical links between witchcraft and crop contaminants such as darnel are both implicit and natural. The infiltration of orthodoxy or states of perceived religious purity by a rogue spirituality, would naturally generate offense.

Symptoms of darnel poisoning will recall those of a more venomous inhabitant of the Spirit Meadow, the ascomycete mold *Claviceps purpurea* or ergot, known to Old Europe as *Mutterkorn*. Commonly known as a parasite on rye, its biology is promiscuous, being known to infest over 600 plant species including barley and winter wheat. Infected spikes of grain appear purplish black and lustrous, and render loaves made from contaminated flour a reddish color. Farming manuals and publications devoted to assuring grain quality have long regarded it as a disastrous pest, and have prescribed strict protocols for detecting and eliminating it. Mass poisonings from inflected grain or flour in the modern era are rare, but known from centuries past.

16 Pearl F. Braude, "Cokkel in Oure Clene Corne: Some Implications of Cain's Sacrifice."

17 Matthew 13:38–40.

DWELLERS OF THE SPIRIT-MEADOW

BOTANICAL	NAME	TOXIC/PSYCHOACTIVE COMPOUND
Claviceps purpurea	Ergot	Ergoline derivatives (LSD &c.)
Agrostemma spp.	Corn Cockle	githagin agrostemmic acid
Lolium spp.	Darnel	loline temuline hydrochloride peramine ergot alkaloids
Bromus spp.	Brome Grass	infection by *C. purpurea*†
Psilocybe semilanceata	Liberty Cap	psilocybin
Melampyrum arvense	Field Cow Wheat	aucubin
Papaver rhoeas	Corn Poppy	rhoeadine

† Several species of brome grasses (*Bromus* spp.) sometimes become ergotized and therefore toxic to humans and livestock.

The affliction of ergot poisoning, known also as Ergotism or Saint Anthony's Fire, takes two manifest forms, convulsive and gangrenous. The former include seizures and psychosis, often accompanied by painful spasm and paresthesia. The latter is characterized by a simultaneous vasoconstriction and edema, the sloughing of skin, loss of sensation, and with the onset of gangrene, tissue-death. Mania and hallucinations can accompany toxicosis, and are the result, in part, of ergot alkaloids and their metabolites. This has led some modern writers to place ergot at the center of certain historical events such as the Salem Witch Trials, though these theories have also been met with sound challenges. In terms of its occult significance, both ergotism pathologies may be considered allied to the general concept of witchcraft, and more precisely, the diabolic or demonic. Additionally, it links with both the witch's art of poisoning and the act of infestation of grain, fact in common with Darnel.

Yet, the maligned ascomycete must also be recognized for its contributions to both twentieth century occult, spiritual and psychedelic culture. The Swiss chemist Albert Hoffman may indeed be considered both exorcist and alchemist, having liberated Lysergic acid diethylamide (LSD-25) from the intermediary ergotamine derived from ergot fungus. In this manner the angelic portions, being the ecstatic and vision-producing principles, were separated from its diabolic attributes, including nausea, seizures, gangrene, spontaneous abortion and death. Not surprisingly, these inverse symptoms of ergot poisoning may be viewed as a microcosm of the vilest attributes of witchcraft and the Sabbat.

A speculative 'Black Bread of the Sabbat' may also have contained the seeds of field-dwelling corn cockles (*Agrostemma* spp.), also known as Crown of the Field. Long an agricultural nuisance, cockle is a contaminant of wheat, soybeans, cereal grains and feed crops. As a by-product of threshing, the seeds of corn cockle emerge, and when they infiltrate grain stores, impart a green or bluish green cast to loaves baked with contaminated flour. Like darnel, seeds of corn cockle are frequently found in ancient archaeological strata such as Roman remains, and earlier.

Agrostemma seeds contain the saponin glycoside githagin as well as agrostemmic acid. These seeds are known to produce a variety of inebriating side effects when ingested, collectively known as 'githagism' and include giddiness, vertigo, delirium, and loss of motor control. They also produce a rancid gastronintestinal efflux, and have also been documented as leading to coma and death by respiratory paralysis. The seed, which contains 5–7% githagin by weight, releases the toxin more readily when cracked, and effects have been reported with the ingestion of as little as one gram.[18] While toxic to humans, the seed is especially hazardous to livestock. Affirming Camporesi's observations, there is evidence that the seeds were used to adulterate cheaper grades of flour in Europe.[19]

As in the case of darnel, industrialized agriculture has taken its toll on wild populations of *Agrostemma*, to the point that in many places where it once ranged broadly it is now deemed extinct. In 2014 a single volunteer of *Agrostemma githago* was discovered at a lighthouse in Sunderland, UK, after having long been thought extinct in the wild due to herbicides and industrialized farming.[20]

Other meadow-dwellers which can become toxic contaminants of grain include multiple species of ragwort (*Senecio*), whose poisoning symptoms in humans lack psychotropic effects but can lead to compromised organ function or failure.[21] Similar multiple organ failure can result from ingestion of meadow saffron (*Colchium atumnale*), due to the presence of the tricyclic alkaloid colchicine.[22] Certain species as agricultural contaminants produce delirium as well as organ damage and death, such as *Trichodesma incanum*. The number of meadow-dwelling contaminant of human grain is considerable, and forage species toxic to

18 N. Kruskal and R. Kobert. 'Agrostemma Githago.'

19 J. R. Mohler, ed. *Journal of the American Veterinary Medical Association*, Volume 10; Volume 57, 1920, pp. 306–307.

20 *The Daily Mail*, June 6, 2014. Varieties of ornamental *Agrostemma* are, however, still grown in private gardens.

21 Interestingly, however, Ragworts have accreted folk names such as 'Fairies' Horse' suggesting otherworldly transport.

22 As with other similar lethal poisons, colchicine shows promise as an anticancer therapy in humans.

livestock are far greater, also inclusive of the complex of fabaceous plants known as locoweeds or crazyweeds.

The concept of the Spirit Meadow mirrors the respective arcana of poison, witchcraft, the Devil, spirits, the Sabbat, the high assembly of the Witches. The atemporal and aspatial qualities which emanate from meadows do parallel some aspects of the Sabbat, as recorded in classic accounts, as well as a state of consciousness that characterizes hypnotic trance itself. This comparison also holds for the particular manner in which, according to folklore, meadows are fecundated with spirit-presences. Linked with the night-wandering spirit-cults of Old Europe as a specific locus of spiritual activity, they also represent a hidden legacy of otherworldly occult power. The entic power of the Spirit Meadow assumes the forms of the angelic, diabolic, or the dominion of faerie, which carries attributes of both. Its most simple emanation is the demarcation and opening of magical locatedness, as well as envigoration of *genius loci*, and bears similarity in some respects to the state attained following the ancient English rite of 'beating the bounds.'

These metaphysical features, as we have seen, are also embodied by some of the flora which inhabit the meadow, linked both to grasses and the contamination of grain crops. Amongst these, the Liberty Cap, which partakes in the highly-charged subterranean strata of the meadow, is a worthy subject occult investigation of these arcana. The fungus *Psilocybe semilanceata*, also called Liberty Cap, is also a meadow dweller, its ecology highly dependent on decaying grass roots in the meadow biome. Clearly the safest of these denizens of the Spirit Meadow, the mushroom is the only one amongst them that has a known history of use within the Sabbatic Tradition, where it was known as 'Crow's Bread.'

The more tenebrous of these grains, though indeed capable of inducing altered states of consciousness, often do so at great physical cost, even to life or limb. Lest the occult fantasist seek to adopt them wholesale as 'visionary allies,' it is well worth examining the bioassays as well as the bills of mortality. Yet their chemistry, co-evolution with grains, as well as their placement within a folkloric context of Sabbats and otherworldly spirit activity, hint at an occult intelligence evocative of the revelations of the immortal gods, and the dark ecstasies experienced at the very border of Death.

THE MATTER OF MAN

IN THE LEXICON OF SABBATIC WITCHCRAFT, *MUMIA* refers to sacrificial offerings from the human temple, living or dead, used as foci of sorcerous power. Such substances are diverse in their origin and usage, but most often encompass blood, flesh, bone and the generative seed of woman and man. Its origin as a word comes in part from the mumia of the medieval alchemist and apothecary—the medicinal preparation compounded from human flesh, most often the pulverized corpse.[1] Not limited to magic and medicine, the putrefactive material of the corpse also achieved apotheosis in the form of the eternal image, as desiccated cadavers of Old Egypt were also ground to render the pigment Mummy Brown, used in oil painting in the 16th and 17th centuries.

1 For additional insight into *mumia* and its related arcana, see Kenneth Grant's discussion of the concept *d-mammu* or the 'effigy of blood' in *Aleister Crowley and the Hidden God.*

The origin of the magical praxis and iconography of sorcerous mumia is rooted in ancient necromancy and the lore of the charnel gods themselves, whose holy medicines include blood of the virgin, fat of the unbaptized infant, and the hand of the convict killed on the gibbet. Inseparable from the unholy figure of the witch, such constituents are to the mundane mind a morbid haunting, steeped in criminal implication, psychosis and the putrescence of Christian heresy. Magical use of human remains was, in the late Middle Ages through the early modern era, often linked with *veneficium*, the art of magical poison. This linkage, frequently on the basis of corpse-desecration or the polluting factors of mortal decay, occurred most frequently in witch-hunting literature and demonological tractates. This criticism also arose from other magical sectors, such as theurgy, astrology, and planetary magic:

> *These Witches and Necromancers who are also called Malefici or Venefici, sorcerers or poisoners; of which names are rightly called, who without the Art of Magick do indeed use the help of the devil himself to do mischief, mixing the powder of dead bodies with other things by help of the devil prepared...*[2]

Apart from its shadowy radices in magical history, the mumic aspects of European witchcraft and necromancy were also extrusions of socio-religious currents already in operation. The culture of morbidity and sanguinous perversions which pervaded European religion, art, medicine, and philosophy has been captured in exceedingly visceral detail by the Italian scholar of micro-history Piero Camporesi.[3] The ravages of bubonic plague in fourteenth century Europe, annihilating a third of her souls, witnessed a spectacle of pestilence, death, and decay scarcely

2 *Heptameron*, attributed to Peter de Abano, 1496. See Also *De præstigiis dæmonum* be Johannes Weyer, a disciple of Agrippa.

3 *Juice of Life*, *Anatomy of the Senses*, et al.

imaginable to the modern mind, and this cultural trauma impressed itself deeply in the collective psyche. Plague doctors, flagellants, and guilds of undertakers and corpse-carriers were surface phenomena of the spectral forces ravaging the land at an invisible level. Secret guilds such as the Antient Order of Bonesmen, whose precise history is obscure, represent a curious nexus of death mysticism, funerary custom, custodianship of the corpse, and occult power.[4]

Yet within this broth of seeming foulness is found not only the wellspring of the sorcerer's power, but of his very life. European folk magic has a long history of mumia cures and as the *materia magica* of spells. In the eighth and ninth centuries alone, no less than eight penitentials imposed strict penance for drinking semen or blood; abundant references reveal a stratum of folk belief utilizing those substances for love and healing magic.[5] Also documented in this period was the practice of women making enchanted medicines from charred human skulls, usually to cure their husbands' impotence. A Bohemian manual of magical cures discovered on a Texas ranch contains the following formula:

> *When you have a toothache, go into an ossuary, take a skull and extract a tooth out of it and rub the gums until they bleed so that the tooth will be covered with blood. Place the tooth back in the skull but not with bare hands and pray the Lord's prayer and the creed once. Say:* as sure as you are dead, so sure it is that my tooth will stop hurting, as nobody hurts you. So help me God. Father, Son, and Holy Spirit. Amen. *But it has to be done at 12 o'clock.*[6]

4 Certain traditions of British witchcraft and folk magic arising in localities near ancient plague-pits also preserve lore related to the corpse and its care.

5 Filotas, *Pagan Survivals*, p. 260.

6 The manuscript is entitled *Perfect and Tested Secrets or Various Medical, Magical, Chemical, Sympathy and Anti-sympathy Treatments.*

This procedure, allied to the rural 'black book' traditions of Germany and Scandinavia, is remarkably similar to some of the charms preserved within extant corpora of present day British cunning-craft. The same manuscript also preserves a recipe to heal wounds caused by weapons using Cranium Moss, thought by some to be of astral or empyrean origin. This moss, said to grow upon certain skulls after long exposure to the elements, was also called Usnea[7], and according to some apothecaries who stocked it, was gathered from the skull of a criminal who had been hung in chains. Apulieus documented the magical power ascribed fingers and noses of persons crucified; the power of corpseflesh (particularly that of a criminal or those who had died violent deaths) for healing, harming and cunning is a widespread feature of folklore. Perhaps the best known such relic is the so-called Hand of Glory, a number of which are preserved in private conservatorships.

Sacrifice, arguably the most ancient sorcerous act, reaches its apex when that which is offered is ourselves –the Matter of Man. When the very substance of our earthly being is given unto the gods, *daimones* and spirits, its mortal function as a vessel is transcended and it becomes a spiritual intercessor. Unlike the Host of the Christian Sacrament, where communion bread must ape the Body of Christ, the witches' mumia, correctly prepared according to traditional magical formulæ, brings the adept into direct spiritual contact with the atavistic retinue, the Assembled Host of the Good, or the companie of Future Flesh.

The value of human material in curing was greatly exploited by the masters of the Alchemical Art, occupying the nether ground between magic, medicine, and chemistry. Sovereign remedies for the preservation of health, of incomparable subtle-

7 Not to be confused with Beard Lichen (genus *Usnea*) which possesses manifold healing properties including antimicrobial and stypic activity. Skull-moss was, nonetheless, an official drug in many pharmacopoeias until the nineteenth century.

ty and power, were distilled from the rudest of skeletal remains. The Alchemists made use of Axungia de Mumia, the corpse-fat known as Marrow of Mummy, Liqua Mumia or human grease, and Liquor Mandibulum, a prized oil extracted from the jawbone. Exalted draughts were also prepared from the blood of the living. This process was perfected by the fire, alembic, and cucurbit, as in such complex formulae as Quintessence of Blood, found in the *Tyrocinium Chymicum* of Beguinus.[8] Implications of an 'alchemie vampyric' are of course obvious, especially in consideration of an incarnate continuity of blood once separated from the human body, and assuming a kind of resurrection in diverse bodies and vessels. At a more important level, the physiologically invigorating aspect of highly-refined human blood also has magical extrapolations, particularly for the energetics of blood sacrifice and the manifestation of sorcery upon the material plane. In this wise, we recall the words of Mercea Eliade:

> *The sum total of vital energy that is left available by the interruption of an existence still far from reaching it natural end is 'creative', in the sense that it is capable of animating any object made by man.*[9]

Though the procedures utilized to formulate such remedies evoked the abominations of the witch's cauldron, they also anticipated the glandular medicaments of organotherapy and modern pharmacy, and, it may be argued, hormone and steroidal therapies. The perfection of human-derived *Azoth* however, lies beyond the flasks and bulbs of the alchemical philosopher, in magico-sexual generation of the medicine by a focused sorcerous application using the human body as the laboratory.[10]

8 Book Three, Chapter 1. The blood must be taken in great quantities from 'sound men in the flower of their youth'.

9 *Zalmoxis: The Vanishing God*, p. 246.

10 See, for example, Crowley, Aleister *De Arte Magica*, passim; Grant, Kenneth

VIII. *Witches roasting an infant, from Guazzo's* Compendium Maleficarum, *1608.*

The mortal body, as the temple of spirit, is also known to sorcery for its strengthening mumic projections into the material world. Possessing the power of highly organized homeostasis, it serves as the building blocks of other bodies, but also of the stone artifices of man. Well-known to archaeology, the foundation-sacrifice and apotropaic house-charm wield the protective power of bone, and its extended potency as a substance of mediation between the worlds of flesh and spirit. This principle is particularly

Nightside of Eden, magico-sexual operations of Temphioth; Andrew Chumbley, *Azoëtia*, emanations of the Lunar Triplicities. The Emanant of Lunafaction, a critical subsection of this arcanum in Sabbatic Witchcraft, is also treated in my own *Ars Philtron*. There are much older exemplars which are religious expressions, such as the medieval Kaula rites; one formula is described in detail in the late medieval *Kāmapānacāstiram* "Treatise on the Arrow of Lust."

important for its implications in esoteric Freemasonry. A similar principle would appear to animate the consecrated stones of the Tahitian *marae,* ritual areas central to the cults of the Society Islands, beneath which human sacrifices were interred.[11]

Ancient deposition of human skulls in wells and sources of sacred waters was a tradition of pre-Roman Britons, as in Carrawburgh—home of a water deity sometimes known by the inscription *Coventina.*[12] Such sacrifices find a curious corollary in the Americas, where the Mayans of old made skull depositions in cenotes, vast underground freshwater caverns. According to tradition, these offerings were made to Chac, the Rain-god.

Cephalomantic augury, utilising skulls and disembodied heads as oracles, was not the sole preserve of the so-called Celtic 'cult of the head', but was also well known to ancient Greek sorcery, and earlier in Mesopotamia.[13] That the practice was established in ancient times is illustrated by a magical operation from the *Papyri Græcae Magicae* whose purpose is to restrain a falsely-prophesying or incoherent skull deemed unsuitable for soothsaying. The mouth of the offending bone was to be sealed with grave-earth; its unseemly predilections further bound with a ring of iron engraved with a headless lion wearing the crown of Isis, and trampling a skeleton.[14] The practice of the oracular skull persists in a number of syncretic sorcerous traditions in modern times; a skull frequently occupies the Ganadero, or sinistral side of the *mesa,* the altar of the Peruvian *curandero,* where it is used as an oracle of the Dead.[15] The parallels between the sorcerous com-

11 Many *marae* to this day remain *tapu* and retain palpably resonant fields of power, centuries later.

12 Merrifield, *The Archæology of Ritual and Magic.* pp 45–46. Also of note are the oil-preserved human skulls incorporated in the fourth-century rubble of the Basilica at Wroxeter.

13 Ogden, *Greek and Roman Necromancy.* p. 208. Consider also the Hebrew Teraphim known to Rabbinic lore, some forms of which were mummified heads, as in the *Targum of Pseudo-Jonathan.*

14 PGM IV. 2125–39.

15 Donald Joralemon and Douglas Sharon, *Sorcery and Shamanism.* Human bones

plex of the curanderos and the traditional witchcrafts of Britain, Europe, and America are striking.

Though the skull is often considered the first throne of sorcerous power, other skeletal remains are attested in conveying the oracular powers of the Holy Dead, including the femur, knuckle and finger-bones, and the scapula. A number of such rites, including the incubation of visions by drinking potions compounded from ale and human ashes, are practiced by the Bonesmen of East Anglia, to whom veneration of the bone is central. Together with their traditional stewardship of graves and boneyards, the Bonesmen possess a sophisticated understanding of the interplay between bone and spirit, the features of which are often shamanic in outline.[16]

The Skull is also held sacred as a magical object in some forms of Wicca, such as Alexandrian Craft, and in extant streams of traditional witchcraft. A modern, exoteric recension of one such ritual venerates the skull as the living oracle of Cain Sa'Ira ('Cain the Hairy'), the Green Man or patron of wilderness, agriculture, and occult herbalism.[17] Though some Craft streams have, as a portion of their sorcerous heredity, come to embrace certain ritual elements of Freemasonry relating to the skull, the inner magical uses serve a far higher station than a mundane symbol of mortality. Use of skull and bones were also well attested among English cunning-folk such as Eggy Roberts, whose hoary cranium-cup was often brought forth in local pubs as a morbid drinking-vessel. This magical artifact, along with other ritual skulls, now resides in the Museum of Witchcraft, Boscastle.

The great power of the *memento mori* has not, historically,

and their fetishistic representations are also employed in the syncretic religions of the Americas including Santeria and Voudon; and in sorcerous traditions such as Brujeria and Palo Mayombe.

16 Pennick, *Secrets of East Anglian Magic*, pp. 62–63.

17 The exoteric recension of this rite is known as 'The Perfumed Skull'; the specific aspect of Cain, *Sa'Ira*, is associated with the *Seirim*, the hoary spirits or djinn of the wilderness.

been the exclusive preserve of folk magicians and sorcerers. The Latin church plied a vast trade in Saints' Relics, including highly-venerated bones, ampullae of blood, and corpses entire of incorruptible flesh. At times, the zealous pursuit of this fascination led to grave-burgling on a massive scale:

> *In the twelfth century the shrine at the cathedral at Cologne obtained the skulls of the Three Wise Men of the East who brought gifts to the infant Jesus. In competition the church of St. Gereon produced the relics of St. Gereon and his whole band of martyrs. The competitive spirit spread to the church of St. Ursula, and a whole cemetery was despoiled to cover the interior walls of the monastery with the relics of St. Ursula and her eleven thousand virgin martyrs.*[18]

We should not be surprised at this example, despite its grotesque nature: in the middle ages necromancy was practiced within the cloister with some frequency, and constituted a kind of "clerical underworld" as magical historian Richard Kiekhefer has observed.[19] Still, the line between good Christian and folk-charmer often lies obscured in the mists of spirit-time, wherein there is a considerable accumulation of heathenism. The water of St. Teilo's well in Llandeilo Llwydiarth, drunk ceremonially from the skull of St. Teilo, a sixth-century Welsh bishop, was of great renown for the healing cures it effected. However the ceremonial draught was presided over by members of the Melchior family, in whose custodianship the skull resided for many years, rather than a priest. Other healing skull-vessels in Wales and Ireland are also attested.[20]

18 Haggard, *Devils, Drugs, and Doctors*, p. 301.

19 *Magic in the Middle Ages*, pp. 153–156.

20 Janet Bord, *Cures and Curses*, pp. 123–124; and Logan, *Irish Country Cures*, p. 42.

Perhaps the most potent usage of osseous mumia lies in the sorcerous veneration of the Mighty Dead and petition for their intercession within the Circle of Art. These remains, reborn as wands, staves, enchanted horns, or simply left in their natural forms, become the unique preceptors of spirit-wisdom accessed by means of atavisms of ghost and flesh, the gate-keepers of the realm of the Dead.[21] Such is the effectuation of bone as material legacy—an incarnative monument raised by the ingenium of the body in accord with lineage, having assimilated the diverse wealth of the earth unto its foundations. Stripped of its putrefactive stages, it is become purified as a new incarnate form, the encharmed bridge betwixt Living and Dead.

Here stands the Monastery of Ancestral Shadow, within whose catacombs the bones of the brethren are conserved by the Faithful as strata of osseous power, the foundations of Atavism that underlie the very Temple of Wiseblood. Bones thus empowered are become the crystallization of the initiatic truths realized by the departed. More than simply facilitating contact with them, they serve the sorcerous praxis of summoning the genius of one's own death, for sacrifice, knowledge, and immediation of power. And if one calls by way of bone, and a voice calls back, it shall be as the creaking wheels of Old Ankhou's grave-cart, come either to stop, or to pass by.

21 Andrew D. Chumbley's *Dragon-Book of Essex* notes that the Sorcerer's Trumpet is made from "a human femur, best stolen from the grave of a saint or a criminal, elsewise taken from the corpse of one's brother in Arte."

THE
WITCHES' SUPPER

THE FIGURE OF THE WITCH IN THE EARLY MODERN ERA was an amalgam of religious typologies including blasphemer, heretic, spiritual malefactor, idolater, consort of fallen angels, and liege of the Devil. In parallel the witch accreted the substance of secular criminality: poisoner, thief, abortionist, grave-robber. These opprobrious brands were impressed on the accused by those whose written records survive, often in the form of legal tractates or penitentials. Yet as command of the printed word spread beyond legal and religious centers, other typologies emerged: healer, folk-charmer, superstitious rustic, impoverished wretch, and others. This procession of witch-guises has continued well into the present day, to include the glamorized images suffused in popular culture: the witch as diabolist caricature, illusion-maker, emanant of sexual allure, and repository of the unexamined *ejecta* of Christian orthodoxy.

An important and little-examined dimension of the witch-guise is that of the reveler at the Devil's Sabbath banquet. The imagery of this feast appears frequently in woodcuts and is occasionally innocuous, but at other times proffers the image of the witch as necrophage. The assembled coven is alternately portrayed as consuming unbaptized infants or the grisly products of desecrated graves; human bones are also included at the table, as they are in portrayals of the witches' Grand Rite. From the perspective of desecration taboo, the array of grim foodstuffs is no less appalling than the relics held in veneration by the Roman and Eastern Orthodox churches: teeth, fingers, jawbones, foreskins and skulls, incorruptible corpses and vials of blood which liquefy and coagulate at auspicious moments. Yet, witches too have their saints and ossuaries, their hallowed relations to the Holy Dead. It is the passage from stewardship and veneration of remains to ritual consumption that triggers affront in the common mind, and has also contributed to the fear of witchcraft. Despite its abhorrent qualities, this forbidden lore persists and is known to some modern practitioners of folk magic as 'The Witches Supper'—a clandestine and disturbing meal which is, in some cases, a cipher for profound spiritual arcana, as well as the lore of poisons.

The process of bodily decomposition was a matter of fascinative obsession and repulsion to our ancient forbears, from both religious and magical perspectives. Upon death, the body naturally undergoes myriad biochemical changes bent toward the singular goal of material retrogression, the descent of the incarnative vessel to the mortified estate of the Profane Adam. Discoloration of tissue, stiffening of the body, abdominal bloating and pooling blood are mere precursors of the great corporeal tumult whose horrific imagery resembles the demonic horrors of the witches' cauldron.[1] Bodily decay produces its own array of chemical poisons, many of which are responsible for the fe-

1 As a dead body decays, a crucial threshold is reached on day three, hence the prevalence of funerary customs allowing the body to lie in state for that period, and a likely factor in the chronology of Christ's Resurrection.

IX. *The Vaudoises Witches,* Le Champion des Dames, *1451.*

X. *Sorcerer preparing cauldron of noxious substances. Tengler,* Der neü Layenspiegel, *1511; courtesy British Museum.*

XI. *Herakles slays the Hydra. German* Schutzblatt, *or apotropaic paper talisman, offering protection against poison.*

XII. *Mandrake image, Richel-Eldermans Collection, Museum of Witchcraft, Boscastle, Cornwall.*

XIV. Mandragora autumnalis *in flower.*

XV. *Tincture of Mandrake, Official British Pharmacopœia.*

XVI. *Witches' Reliquary for Mandrake Alraun, from the collection of the author.*

XVII. *Sorcerers conjuring spirits in magic circles, shown with horn-thurible for suffumigation. From* Hollenzwang *grimoire corpus.*

XVIII. *Corn Cockle (Agrostemma githago).*

XIX. *Autosentience of the Venom-I, by Andrew D. Chumbley.*

tor so viscerally offensive to the living nose, but also serving as inviting beacons to scavengers and detritivores. The fortress of primordial Adamas, once inviolable with God-given dominion over Nature, is rapidly transformed into a food source for a great variety of organisms, this status heralded by the production of corpse-poisons. Many of these putrefaction-derived compounds, in isolation, can be intoxicating or deadly to *Homo sapiens*;[2] some of them, in minute amounts, are also associated with pleasure or sexual allure, thereby recalling the ancient connubium between Eros and Thanatos. In some cases the corpse-poison also served a magical function before physical death: the power to cause flesh to rot on a living body, by forced infection and corrupt magical principles, was a known power of Zuñi medicine men and a documented procedure during the slow execution of witches.[3] This odorous stew of nitrogenous cadaver-compounds falls into the ancient toxicological classification of ptomaines, from the Greek *ptōma,* indicating a corpse or "fallen body." Their provenance is the graveyard and charnel house, the crypt and plague-pit, and they are united in both science and magic as the vaporous effluent of the necropolis.

. . .

2 Such as the 'blue haze', halo effect, and other visual distortions accompanying exposure to critical levels of triethylamine, a product of organic decay.

3 Simmons, *Witchcraft in the Southwest,* p. 111.

TABLE 2: *SELECTED CORPSE POISONS*

COMPOUND	CHEMICAL FORMULA	SOURCE
Putrescine	$NH_2 (CH_2)_4NH_2$	Decaying Animal Tissue
Cadaverine	$NH_2 (CH_2)_5NH_2$	Decaying Animal Tissue
Trimethylamine	$N(CH_3)_3$	Decaying Animal Tissue Fish *Crataegus* species
Triethylamine	$C_6H_{15}N$	Decaying Animal Tissue Dog's Mercury
Skatole	C_9H_9N	Intestinal Putrefaction Microorganism death Feces Coal Tar Neroli and Jasmine
Methanethiol	CH_3SH	Organic decomposition Feces Coal Tar Swamps Asparagus metabolite
Pyridines	base C_5H_5N	Bone Oil Vaginal secretions *Atropa belladonna* (root) *Althea* spp.
Spermidine	$C_7H_{19}N_3$	Putrefaction of Animal Tissue Semen *Citrus* x *paradisi*
Muscarine	$C_9H_{20}NO_2+$	Decaying Animal Tissue

No less than the natural decomposition of the human body, foodborne illness is also caused by organic decomposition, and has been colloquially referred to today as a kind of poisoning. Corrupted food been a perpetual fact of civilized existence and has required ingenious solutions to forestall the advance of decay. Transmitted by the noisome taint of worms and micro-organisms *en masse*, putrefaction was a philosophically confounding process both dead and alive; the stench and ugliness generated in contaminated victuals were likewise an offense to reason as well as the senses. Early technologies of food preservation included cooling, drying and salting to arrest decomposition, or, in some cases, to mask objectionable flavors of rancidification.[4] The ancient arts of meat preservation naturally share a kinship with embalming: the outrage of post-mortem decay was of prime importance to the Old Egyptians, whose methods of providing salvific respite for the corpse may rightly be considered a magico-religious art form. In Christianity, the processes of corporeal decay were assigned to the dominion of the Devil, likely one reason for the folklore that Satan cannot abide the presence of salt. Persons who claimed to have attended the medieval Witches Sabbat remarked on the absence of salt at the feast. Similarly, when salt was brought in, the spectral revelers of midnight's table suddenly vanished, leaving the guest alone. The power of salt for slowing or arresting decay also relates to its magical uses for exorcism, blessing and consecration. The magician's exorcised circle is thus both fortified and mummified, a perfectly-preserved moment in time and space.

Both the corrupted products of Death and the means of slowing or arresting them bear crucial relationships to the Witches' Supper, which in one interpretation (stripped of its heretical elements) can be seen as fostering a ritual intimacy with the de-

4 Spices such as Juniper and Ginger played an important role for overcoming the effluvia of rotten meat, as with medieval European cooking, which bears certain parallels to the Witches Supper.

ceased. That the witches' delectations should be portrayed in the first instance as necro-cannibalistic is consistent with the position of witchcraft as transgressive, and as operating in spheres roundly condemned by religious and social orthodoxy. The witches' relation to the dead vis-à-vis their atrocious meal is, on the surface, portrayed as a mock Christian communion, or as the vulgar tactic of demonizing enemies by implied cannibalism. On a different level, the Supper operates as a hieroglyph of specific witchcraft power, namely the unique magical relationship between witches and the so-called 'Mighty Dead', the retinue of ancestral shades and fountain of pre-incarnate atavism. The art of necromancy, or magically calling forth the shades of the dead, has long been a vibrant strand of witchcraft and magic of many epochs, and in many recensions may be considered its driving engine. Linked with more ancient currents of shamanism, this art was known from the writings of ancient Sumer, Chaldæa, and Greece, the latter providing the prototypal witch-figure and poisoner Circe, the sorceress of Homer's Odyssey.

The materia of the dead—flesh, blood, and bones—is the mumia of art, known well to witchcraft, alchemy, folk magic, and medicine.[5] The act of its ritual consumption, presented in early modern Witches' Supper depictions as vulgar cannibalism, encodes a number of precise ritual formulae and powers in necromantic magic. The most important of these is the elevation of 'dead' matter to a living state by its incorporation into the living body. This is the active principle underlying the Holy Eucharist, wherein, through divine transmutation of elements symbolizing the mumia, Christ's body and blood are come forth from the tomb, and commune with the body of the Faithful. The potent

5 Pulverized corpses were for centuries an essential part of the official preparations of the apothecary and pharmacist. After falling out of favor for a short time, the ancient wisdom has been adopted again in the form of medical transplant technologies. A common usage in the present era is the bone powder derived from donor cadavers for use in bone grafting and dental implants.

necromantic implications of the Holy Communion, as a magical act, would have been instantly recognizable to practitioners of folk-sorcery, particularly in contexts where funerary rites maintained close communication with the departing spirit.

Present within the feast of the Dead is also the Formula of Opposition, a precept which underlies many historical patternings of witchcraft. Named by Andrew Chumbley, who wrote about it extensively,[6] the Formula is an operant dynamic between the sorcerer and the 'Other', that being the zones of spirit-alienation external to personal experience and containing ungathered seeds of occult numen. In the case of historical folk magic, Formulae of Opposition are often transgressive against law, religious orthodoxy, or social convention, but above all against Self; as exacted they often make use of inversion.[7] In violation of strongly-held personal tabu, the structure normally governing conception and use of magical power is overturned, resulting in a liberation of consciousness, and the acquisition of previously-forbidden realms of power.[8] At the Feast of the Witches, a culinary encounter with dismembered limbs, organs, and heads serves as an oppositional force on a multitude of levels, from the basic violation of the senses, to affronts against personal and group morality. Whilst the actual consumption of decomposing human flesh by historical practitioners of Sabbatic rites is an open question, it is, perhaps, the wrong question. More relevant is the depictions of the moribund feast as a symbol of initiatic power gained through the Formula of Opposition.

6 *Azoëtia, A Grimoire of the Sabbatic Craft (1992) Qutub* (1995), and *The Dragon Book of Essex* (1997) are exemplary in this regard; see also my own *Lux Haeresis* (2011) which treats of the matter as active upon the Path and its perception.

7 For example, spells using stolen goods, utterance of the Lord's Prayer backward for gaining diabolical power, the 'profanation of the Host' etc.

8 When manifest in actuality, the Formula acts as both a rejuvenating force and as a sublime mechanism of spiritual equilibrium. However the Formula is often misconstrued by those seeking a dogma to excuse sociopathy and a pretense of the 'diabolic.'

XX. *Witches and diabolic consorts at the Sabbat Feast.*

The Accursed Victual, as a component of the Feast, may also mask the presence of initiatic power, conveyed through mumia. A recurrent component of magical charms is the secretion of semen, menstrual blood, feces, or urine into food as a spell of control over one's victim. This action mimics the spoor secreted by many mammals for the 'marking' or 'claiming' of territory and if correctly engaged draws upon a vast astral repository of atavism, and belongs to an ancient stratum of magic reaching into prehistory. Spells employing such secreted matter are transgressive of ancient dietary laws wherein food, and the feast itself, represents a sacrosanct compact between the dining parties. However, when the parties are wholly conscious of the nature of their food, and eat nonetheless—as they are shown doing in portrayals of the Witches' Supper—it may be presumed that there are religious or magical reasons for doing so, namely reverence for the deceased, the acquisition of power, or both.

All such approaches to the Feast are essentially necromantic, and as a coercive approach to spirits, it is properly classed as sorcery. It is thus aligned with early modern witchcraft, but ritual communion with the dead using food and drink is also a feature of ancient religion.[9] Roman cults of the dead persisted into the early centuries of Christianity, with night-long memorial feasts in honor of those whose bodies had passed, often in situ at the tombs themselves. Archaeological evidence, as well as the written record, reveals remains of ancient graveside banquets, including drinking and cooking vessels. Church prohibitions on pagan rites honoring the dead occurs in written form as late as the thirteenth century, indicating that such observances were still in practice.[10] Feasts offered in honor of the dead persist into the modern era, even in exemplars largely bereft of religious trappings. Ritual consumption of the dead as part of a socially acceptable funerary practice, is also documented.[11]

The abominable meats, bones, and sundered limbs often pictured at the Witches' Supper may be afforded an additional interpretation with regard to their magical rôle at the Witches Sabbath. In certain inquisitional records, an emergent pattern among some groups, which differed from the usual clerical projections, involved a banquet with archaic features which scholar Wolfgang Behringer has called 'The Miracle of the Bones.'[12] This features the restoration of life to a cow or other animal from a disjointed skeleton. The implicit power of this mystery as a magical practice is captured in a section of Robert Fitzgerald's *Midnight's Table*, a manual of witchcraft lore and spellcraft concerning the arcane mystery of the witches' banquet:

9 Lucian reports the use of the pipe for grave libations; Hebrews planted trees on famous graves and made libations there; known as *mazzebôth*.

10 Ramsay MacMullen, *Christianity & Paganism in the Fourth to Eighth Centuries.*

11 As with the Aghori, Yąnomamö, and other groups.

12 *Shaman of Oberstdorf.*

The Mind void yet the Thought fully formed.
The Body hungry yet the Spirit replenished.
The Wood unfinished yet the Table carved.
The Platter empty yet the Larder full.[13]

Here the desolation of the witches' feast remains, as well as their potentiality as nutritive victuals or even as living beings, is invoked, the suggestion of Voidful Presence through the juxtaposition of emptiness and corporeal flesh. Extrapolated beyond the objects themselves, the table may be seen as the witches' altar or circle, the zeroth vessel of all-potentiality which, like a cornucopia, may contain a multitude of fruits by way of ritual power. This symbolic and emblematic patterning is completely consistent with the atavistic patterning evident in the orally-transmitted magical lore of the Sabbatic Cultus.

The natural transformative processes of rot and decay are crucial strands of the magical currents feeding folk magic and witchcraft. The alchemists of Europe explored putrefactive states thoroughly, borrowing the process from Nature, then emulating, calibrating, and magnifying it under precise fractionations in glass vessels. It is likely that, as with the Royal Art itself, a considerable portion of putrefactive magic in Europe was a direct inheritance of Arabic and Islamic magic; such texts as *Ġāyat al-Ḥakīm* and *Kitab al-Sumum* employ numerous members of dead animals, some ritually killed, for cursing, poison, and magical power. These usages also occur in the later corpus of European grimoire formulæ. However, the powers of putrefaction and decomposition had a far more ancient pedigree, one of which is of specific interest to the Sabbath banquet. Correctly harnessed, they give rise to both of the primary mysteries of the witch sacrament: the bread and wine.

In the Bread and Wine of the Witches' Supper, some have

13 'Tabula Rasa Sabbaticum,' *Midnight's Table*. Privately circulated, the book's publication is forthcoming through Xoanon Publishing.

seen the historical outlines of the ritual consumption of psychoactive substances at the Sabbat, specifically conveyed through food and drink, and indeed this interpretation is present in some modern-day witchcraft practices. Historical references are uncommon, but suggestive. The Inquisitor Pierre DeLancre reported that the bread of the Basque witches was black and revolting, its flour ground from black millet. Aside from its resemblance to cadaverous flesh, the 'black bread' is of potential toxicological interest. In centuries past, white flour was a privilege of the wealthy, and poorer classes resorted to eating so-called 'black breads', made of rye and barley, and which also contained all manner of adulterants from the harvest. Piero Camporesi in his *Bread of Dreams* has speculated that psychoactive contaminants of grain such as darnel (*Lolium temulentum*) and ergot (*Claviceps purpurea*) were so common in the flours of some regions and eras that the average peasant was in a constant state of intoxication. If true, the evidence cited by Camporesi suggests that the psychoactivity of such breads was an accidental by-product of a fouled food supply, but if the phenomenon was understood by herbalists and magical practitioners, there would be little to stop the cunning from crafting experimental loaves.[14] Indeed, as with the Thelemic 'Cakes of Light' the Sabbat-Bread has its own secret formulations.

The old term "Crow's Bread" originates in the founding lineages of the witchcraft order Cultus Sabbati, and originally referred to the intoxicating mushroom *Psilocybe semilanceata* as a gift of the spirits for visionary ritual use. In the late 20th century, the term was applied within the group for broader use to refer to any psychoactive ritual substance gathered from Nature, but

14 The obvious problem in formulation is that the biochemically active adulterants of such speculative breads carry considerable risk of undesirable pathologies, such as gangrene, necrosis, and convulsions brought on by ergotism. Additionally, human bones can harbor harmful bacteria such as tuberculosis long after physical death.

its nature as 'Bread' is linked both with the Communion Host of Christ and the male generative power linked with the 'Lord of Light', in some cases identified with Lucifer. In this latter association, the Bread's power as Revelator is especially notable. Covines and lodges of the Cultus have long made use of venefic gnosis in various forms; its oldest known recensions, dating from the second half of the nineteenth century, contain obscure charms against poison, as well as certain ritual transmissions of power using a prepared psychoactive sacrament.[15] Oral teachings long pre-dating the Great War concern another poisonous species of note in Britain: Belladonna. There are also adjunctive practices concerning a multitude of other plants of power, specifically their Eucharistic power.[16] My contacts with other traditional witchcraft groups outside of the Cultus have, on occasion, affirmed the presence of such sacraments elsewhere, some of which have themselves passed into a largely symbolic or chemically inert form.

Within the modern Sabbatic Current, the Bread of the Sabbat Feast operates upon many magical levels, its essence is intimately tied to British agricultural cycle, the God of Harvest, Corn and Sheaf, sometimes manifest in the mythical divinity of John Barleycorn. The germ of this myth encloses the great mystery of ritual murder and resurrection embodied in the Holy Loaf, and the resulting sustenance of the kingdom. This quintessentially English expression of the Bread is thus seminal, nutritive, life-giving, and radiant, but also embracing the mysterium of Death. In the Sabbatic Witchcraft tradition, Barleycorn is sometimes identified as the Witch-Father Mahazhæl. He is thus often de-

15 Among many adepts, attitudes toward drug-taking outside the circle or the crucible of the Grand Sabbat are sharply conservative; regarded either as the decadence of thrill-seeking counterculture, or as a base pathology.

16 Chumbley, A. "The Golden Chain and the Lonely Road", *Opuscula Magica Vol. 1.* Ritual use of psychoactive substances was also an interest of the traditional witch Robert Cochrane (1931–1966); see Monmouth, John: *Genuine Witchcraft is Explained.*

picted as a skeletal god with an erect phallus, bearing a scythe, sickle, and stalk of grain; his mystery is well encapsulated in his invocation from Chumbley's *The Dragon-Book of Essex*:

On the first day I awoke within the furrow.
On the second day I knelt in prayer 'neath the sun.
On the third day I stood in the long green robe.
On the fourth day my head was crowned with gold.
On the fifth day the sickle laid me to rest.
On the sixth day my body was ground between stone.
On the seventh day I was raised anew
to feed the brethren at Midnight's table—
to serve at the Round Feast for both the Living and the Dead.[17]

In addition to the process of ritual murder which births the Bread, the putrefactive processes used for its fermentation, via yeast or bacteria, are also reckoned as a part of the Corn-God's dominion. As a natural agent of corruption, yeasts are widespread and penetrate countless strata of the world, often contaminating foodstuffs, as well as the human organism. Even where fermentation conditions are controlled, the process of making bread and wine relies on the mass death of these microorganisms. This catastrophic loss of life, on the order of hundreds of millions of individuals per loaf, nonetheless provides a delectable crumb serving as both holy sacrament and the common man's 'Staff of Life'. A further relation between bread and the grave is its frequent off-white color, recalling bone, and the hardness it attains when stale, sometimes petrifying, as a skeleton, over the course of centuries; and amongst some witchcraft practitioners, the churchly Communion Wafer is sometimes addressed within the circle simply as 'The Corpse' or 'The Skeleton.'

17 "Invocation of Mahazhael-Deval, Our Lord of Midnight", from the Rite of BHA, *Dragon-Book of Essex*. This was also excerpted in the witchcraft and folklore journal *The Cauldron*.

XI. *The Nocturnal Assembly gathering corpses for the Witches' Supper.* Compendium Maleficarum, *1608.*

The magical corollary to the Witches' Bread is the *Vinum Sabbati,* or winecup of Midnight's Table. Its alignment is with the Moon and the Lunar emanation, the feminine principle, and the many humors of the body, primarily blood, but also the female sexual secretions, both gross and subtle. In witchcraft contexts, as well as other secret societies and magical orders, the Wine is of legendary status and a great deal of lore and doctrines have emerged concerning its generation and use. To some it is a cup producing fantastic visions, to others, an initiatic ordeal which serves as the most harrowing trial for the drinker. Certain teachings, through its association with both the Living Cup and its Wine as a single entity, have two essential natures which in combination, magically unify to create a Blessed Third, an apotheosis of both. Within the Cultus Sabbati, the 'Graal of Midnight' has precise formulations to empower and support the various pathways of Sabbatic Congressus: Thanatomantic, Atavistic, Sexual, and many others. By a metaphoric pathway, the Wine of the Sabbat is not only a fluidic medium, fermented and distilled within the Flesh of the Initiate, but also the entire process of corporeal transmutation during its imbibition at the High Sabbat.

As an actual drink conveying ritual power, a medieval prototype of the Wine of the Sabbat is to be found in Johannes Nider's *Formicarius* (1435), which alleged the witches of the Simmenthal region of Switzerland were initiated using a potion brewed from the ashes of infants. More important than the composition of the brew was its alleged effect: the beguiling draught conferred upon the initiate an instant knowledge of the Art Magical. Though described prior to the advent of the Sabbat as a major component of witchcraft, it is the ritual cup and its function as a bestower of witch-power which links it to the Witches' Supper.

The bridge between wine and the incorporeal host is also relevant to the nature of the witches' cup. Historically, the grape was considered divine not only by mankind but also by spirits of the Dead.[18] In ancient Greece, the Vine-shoot was regarded as possessing strong properties of purification; wine was often poured there as a libation for the dead, as well as to chthonic deities. This custom of offering alcohol to the deceased resisted the strongest attempts at eradication; Theodoret, Bishop of Cyrrhus (427–449) reports with outrage pagans bringing wine to the deceased in evening rites. Cæsarius relays a legend in which two servants at the monastery of Laach, charged to guard the vineyard by night, bribed the devil to do their work with a cophinum full of grapes, a deal which was apparently kept. Amongst the nocturnal activities later alleged of the Vaudois witches was the

18 Let it be remembered that the first deities of the Vine were goddesses; this was to change in Egypt, where Osiris achieved primacy. As lord of the Dead, he was appeased with vinous libations. Wine was also used as nourishment while journeying in the Spirit Realms.

invasion of wine cellars, led there in a troupe by the Devil. "Under lead of the demon they enter cellars and drink wine, all of them first urinating in the cask from which it is drawn."[19] The threefold linkage of wine to the Dead, witches and the Devil draw additional lines of arcane association with the Sabbatic Grail, both as a form of communion with the Dead and with the Black Man of the Sabbat.

The presence of Wine in historical English witchcraft and folk magic may indeed arise from its aspect as mock-sacrament, the 'polluted blood of Christ' which featured in invertive and blasphemous sorceries. However, wine was present in England before the advent of Christianity; introduced by the Romans, there is evidence for viticulture among the Anglo-Saxons; one conservative estimate identifies at least 139 definite or possible vineyards in medieval Britain.[20] Though climatological trends in past centuries have fluctuated, and viticulture has prospered or suffered accordingly, the Genius of the Vine has been present in England for millennia. This is certainly sufficient time for a body of lore and rites to have accreted around the grape and its divine expressions, drawing from numerous magico-religious currents, as well as the inevitable corpus of agrarian lore which accompanies so important and venerated a crop. This is to say nothing of England's great tradition of hedge wines, a testament both to the ingenuity of her vintners and the botanical diversity of her lands.

The Cup of Wine which features so prominently at the Feast of the Witches may be understood as the mechanism of sorcerous transmutation of the body, not only its vehicle, but its sym-

19 Alphonso de Spina, *Fortalicium Fidei*, 1494. "Men lying awake in bed will often hear someone walking and breaking things with heavy blows, especially on wine casks..."

20 William Younger, *Gods, Men, and Wine*, pp. 237–39. Often dismissed as a wine producer, England produces many outstanding wines, such as the emergent market of English sparklers, noted for their resemblance in product and *terroir* to the famed Champagne of France.

bol, process, teaching, and legacy. This symbol in activated form unfolds, as an opening rose, the entire ecstatic algorithm of the Sabbat. Within the rites of Sabbatic Witchcraft, the Wine of the Devil's Graal appears in radiance at the confluence of sorcerous enchantment and spirit-veneration. Where the covenant of adepts is of sufficiently focused will, desire, and belief and of sincere devotion[21], the Cup is vinted, filled, mixed, and drunk. The motto *Ipse venena bibas*[22] or 'drink thine own poison' encodes the truth that the Grail of the Witch is both the cup from which it is drunk, and the initiate into whom the wine passes. The alphanumeric essence of this matter is eloquently contained within the number 710, which corresponds both to the grail-poison (*tar`elah*) and the Sabbath itself.[23]

The active magical nature of the Witching Graal, and its function as the intermediary in rites of 'Communion' naturally evokes the Body of the Goddess as the portal of mystery. Kenneth Grant has linked the Sabbatic Wine to the blood of Charis, wife of the smith-god Haephestos, and also known as the threefold goddess Charites, or the Graces. Expanding upon the writings of Massey, which quote the ancient writings of the Gnostic Marcus, Grant links the Vinum Sabbati with the blood of Charis, the 'original Eucharist' of the early Gnostic Christians. The vintage is the central component of the ancient magico-sexual rites of trance mediumship wherein the goddess spoke through a chosen medium.[24] This bears certain similarities with kindred operations in the Order of Eastern Templars, as well as those of at least one tra-

21 The former is linked with magic, the latter with religion. However, in the case of Sabbatic Witchcraft, spirit-veneration bears only a remote kinship to religion, as, for instance, dogma is replaced by tabu.

22 From the Catholic exorcistic formula *Vade retro satana* or 'step back, Satan', certain forms of which are known as traditional witchcraft charms.

23 Biblical instances of *tar'elah* are scarce, but see Psalm 60:63 for the *yayin tar'elah* or 'wine of trembling or staggering'; also in Isaiah 51:17 and 51:22, it is 'The Chalice of Reeling.'

24 Grant, "Vinum Sabbati", *Carfax Monographs*. The Wine of the Sabbath was also mentioned at length by Arthur Machen in his 1895 novel *The Three Impostors.*

ditional witchcraft lineage informing Cultus Sabbati. Likewise, a cup-blessing used for the Wine connects its use to the forgotten intimacy of Samael and First Woman:

Bright Host of Saint Hawa, draw nigh unto this, my Cup.
Before mine eyes, the Well of Abomination,
Betwixt thy thighs, the Red Stream of Eternal Fire.
Behold thou the Good Companie assembled
To feast upon the grave-wandering corpse,
Draught of Manbane, and dew of the Forest grail,
The blood-fouling thorn, the Fang and Toad-froth,
All Delights of Resurrection's Vineyard:
O, Mercy of the Spirit I pray![25]

Here 'communion' also relates in mystery both to the Witches' Agapae or love-feast as well as the coition of spirit transpiring within the circle of the High Sabbat itself. This resonates with the witches' *Fortunum* or Cup of Good Fortune, a specific preparation of male and female sexual secretions, ritually expressed in the correct lunar phase, and empowered through conjuration of precise spirit-presences. Within these covines are preserved teachings concerning 'the vinting and pouring' of the Agapae-wine, as well as its function at the Feast. It is impossible to pinpoint with certainty the origin of the oldest of these witch-rites, though their resemblance to some practices of South Asian tantra is striking.[26] This may be an occult adaptation of Tantric practice, as perpetuated through such magical orders as the Ordo Templi Orientis, with which some covines have had contact. However, the oldest witch-praxes of this type pre-date the Oriental Templars' contact with Tantra, and in fact retain elements marking their origin as specifically English and Northern European. Additionally, their foci incorporate atavistic formulæ,

25 From a private ritual manuscript from a cell of the Cultus Sabbati.
26 See David Gordon White. *Kiss of the Yogini.*

placing them squarely within the precincts of an ancestral cult, as well as incorporating elements which would to many occult lodges be considered 'low magic.'

Despite the linkage of these sexual witchcraft formulae with the Dead, their strata of magical expression very much concern the living, the present body of initiates, woven into the perpetuity of magical time. In addition to the powers of manifestation their perfected exaction radiates, they are capable of simultaneous intoxication, empowerment and nourishment – the great 'Transmutation of the Body' in which one becomes magic entire. Its linkage with the ghastly imagery of the demonologist lies in its formulation from the *Corpora Humanis*. Under correct conditions, the two give rise, like the antediluvian pillars, to the Great Temple of the New Flesh.

Returning to the concept of Crow's Bread, within the Sabbatic Cultus the Liberty Cap mushroom (*Psilocybe semilanceata*), when encountered growing in the wild, is regarded as an omen of ancestral favor. A prime concentrator of atavistic force, it is a gateway to the dominion of Faerie and a guardian of the Way. It is never hunted, but when encountered must be acknowledged by certain ritual customs and sacrifices.

Importantly, it eschews dung, unlike other visionary mushrooms of its genus, and thus in mystical terms is separated from Abel, the unrefined or 'profane' nature of flesh prefiguring the sorcerer Cain. Proceeding as it does from the soil and thus the subterranean vaults of the Mighty Dead, its fruiting body is the brief apotheosis of those fallen and yet come again: the ephemeral Risen Phallus of the Spirit-Meadow. The mushroom thus subsumes three important mysteries of the Witches' Supper in one body: the Corpse, the Phallus, and the Visionary Sacrament. From a devotional entry in *Hypnotikon*:

> *Amongst the true-born of its flesh, it is known as 'The Watcher on the Moor' and this is precisely where I was introduced to this Friend. It speaks of many things: great spectral mists uncurl-*

> *ing before the moon; of time and the procession of bodies upon bodies; of hedge-haunting devils; of the deeds of the Saints' bones, resonant and deep in the earth; of the Immovable Stone and its wisdom; of symmetries and arrangements of things—trees, plants, beasts; of holy books writ in ossuary dust; of the delectations and radiances of the flesh; of the Round Dance and the Fallen Star; of the Sovereign and Horn'd Head detached from the body, ruling over the Land; of the telescoping of the soul into indescribable abysses. When it has spoken its final word, and revealed its last vision, what then remains? The accumulated counsel of every incarnation as 'I.'*

In the abyssal heart of ancestral shadow, the 'Bread' of Midnight's Table is served *both* for the Living and Dead. For those who sup in flesh, and walk in the world of men, it is a sacred loaf broken for remembrance: to honor the Dead with sensation and savor, and to call forth into the body, through the rite of *necrodeipnon*, what has gone before. For them who abide in shade, the Bread is the Lantern of the World, shone as a beacon for return to the flesh, if ever briefly. Through the medium of poison, and its child ecstasy, the decay and annihilation of Death is cast aside, the spirit clothed anew in the radiance of corporeal transfiguration.

INFERNAL VAPOURS

THE FORMULARIES OF THE CONTINENTAL GRIMOIRE OR 'black book' harbor a corpus of incense recipes unique to the magic of spirit-conjuration, and of preeminent significance in the traditions of occult botany. Texts such as *The Sworn Book of Honorius, The Sacred Book of Abramelin the Mage,* and *The Grand Grimoire* are classic examples of the genre and include, among the magician's needful regalia, the brazier and incense, in order to facilitate conjuration of angels, demons, and helper spirits. Plants comprising these incense or suffumigation formulae are of interest to the toxicologist as well as the practitioner, as are their powers of incepting preternatural states in the sensorium. Likewise, a number of allied incense recipes appear in other accounts of sorcerous practice, where poisonous plants play an important rôle.

Incense has a very ancient pedigree, and its use in sorcery and religion makes its appearance in the earliest writings. While the perfumed offering has enjoyed magical uses as diverse as purification, protection from demons, invisibility, storm-raising, and other thaumaturgic aims, the two most important powers attributed the suffumigant are as a votive or sacrificial offering to spirits or god (most often the concern of cultic religious practice), and as a means of summoning and commanding spirits (the work of sorcery). In addition to these, we may also include the use of incense as an aid to meditative and visionary states, which is present both in religion and sorcery, but may also exist as a secular pursuit.

Considering the concealed nature of sorcery, as well as historical cycles of the mass destruction magical books, it is certain a vast number of ancient formulae for conscious-ness altering magical incenses are now lost. Despite this many exemplars are preserved in written form, containing plant constituents with known psychoactive properties. Herodotus spoke of an inebriating rite of the Massagetan peoples, who, due to their proximity to the Scythians, probably employed cannabis: the 'fruit' of a certain plant was thrown onto coals, and the resulting smoke provoked intoxication, song and dance.[1] potent suffumigant is also prescribed in the *Sword of Dardanos*, a Græco-Egyptian love spell summoning demons under the auspices of Eros and Aphrodite to 'bend and attract the soul of whomever you wish':

> *The burnt offering/which endows Eros and the whole procedure with soul is this: manna, 4 drams; storax, 4 drams; opium, 4 drams; myrrh, [4 drams]; frankincense, saffron, bdella,/one-half dram each. Mix in rich dried fig and blend everything in equal parts with fragrant wine, and use it for the performance. In the performance first make a burnt offering and/use it in this way.*[2]

1 As noted in *Pharmako/Poeia*, at least two Russian archaeological finds substantiate ceremonial use of cannabis among the ancient Scythians (pp. 184–185).

2 BPGM IV. 1716–1870. See Betz, *The Greek Magical Papyri in Translation.*

The Sword of Dardanos incense composition is notable for the presence of opium, known to produce expanded sensoria when the smoldering gum is inhaled; it is also categorically hypnotic, and generative of fantastic reverie. When burned, Saffron is a putative psychoactive and is present in numerous medieval and early modern incense recipes; it is medicinally considered an exhilarant and aphrodisiac, which when orally ingested to excess, has narcotic effects. Frankincense, generally ignored in popular entheogenic literature and investigations, is also significant. Recent scientific investigation has provided physiological confirmation of what sorcerers, ecstatics, and religious devotees have known experientially for millennia: inhaled Frankincense smoke is psychoactive.[3] The confluence of Opium, Saffron, and Frankincense vapor in an enclosed working space, together with the repetition of mantic utterances and the force of singular magical focus, may be considered the sorcerous matrix *par excellence* for manifestation of the spirit. Indeed, the words of the incantation itself are adumbrative of the plant-gods in the incense formula, as well as their effects. The Daimon is described as *most headstrong, lawless, implacable, inexorable, invisible, bodiless, generator of frenzy, archer, torch-carrier, master of all living/sensation and of everything clandestine, dispenser of forgetfulness, creator of silence, through whom the light and to whom the light travels.*

Other ancient incenses used to contact the spirit world possessed psychoactive properties. Siberian Tungus shamans burned wild rosemary (*Ledum palustre*) some-times mixed with juniper, for trance induction; the plant was likewise used by the Giljacs for oracular work. The plant's strong inebriating properties are capable of causing 'frenzy and deliria.'[4] Ledum's use as a traditional trance-inducing and calming incense has been noted

3 Arieh Moussaieff, et al. "Incensole acetate, an incense component, elicits psychoactivity by activating TRPV3 channels in the brain." *FASEB Journal*, 22, 2008.

4 Rätsch, *Dictionary of Sacred and Magical Plants*, pp. 187–188. Genus *Ledum* is now classed as *Rhododenron* by most taxonomists.

by Fischer-Rizzi in her modern incense formulary, which provides a recipe using the plant, compounded with 1 part chopped, dried ledum; 4 parts pine bark and resin; ½ part hops; 1 part juniper berries and 2 parts frankincense.[5] Related *Rhododenron* species also possess similar properties. The modern ecstatic rites of the Hunza *bitan*, wherein the ways are opened to communicate with the ferocious spirits known as *pari*, are assisted in no small part by the smoldering branches of *Juniperus macropoda*.[6]

Though irresistible to spirits, ancient magical perfumes also command the opposing powers of this axis: that of chasing, or expelling spirits from the arena of occult operations. This magical knowledge is widespread and interpenetrates the magical traditions of many locales and eras. We find written examples of this in antiquity such as that mentioned in *The Testament of Solomon* (c. 3rd century AD), in which the venerable magician 'frustrates' the arch-demon Asmodeus with the smoke of liver, fish gall, and a storax branch.[7]

In more recent times, wild cherry wood has been smoldered as a fumigant by Greek peasants to drive away the *Kallikantzaroi*, the rapacious theriomorphic beings of the underworld known to walk the earth during the Ember Nights, and who wander the earth at noon and midnight. Cherry and its arboreal kin (*Prunus* spp.) contain varying amounts of hydrogen cyanide (HCN) or Prussic acid in the leaves and kernels.[8] The psychoactivity of the

5 *The Complete Book of Incense.*

6 See M. H. Sidky, "Shamans and Mountain Spirits in Hunza." *Juniperus* species are used throughout the Himalayas in traditional incenses, as are the leaves and twigs of the toxicologically important *Rhododendron anthopogon* and *R. setosum*, with which they are sometimes mixed. The magical significance of the goat as an intercessor of the *pari* is also notable.

7 *The Testament of Solomon*, 6:4. J. Priest, trans. (*Old Testament Pseudepigrapha*, Vol. 1, 1983). Though preceding the main corpus of European grimoires by many centuries, the Testament is in many ways an essential part of their ontological foundation, specifically with reference to the legend of Solomon commanding demons by means of Angelic power.

8 Hydrocyanic acid was an agent of human extermination employed in Nazi con-

inhaled burning products of *Prunus* species are touched upon by Pendell in *Pharmako/poeia*, specifically with reference to *Prunus emarginata*.

The corpus of European suffumigations burnt to purify the airs during outbreak of plague constitute a significant stratum of this art, with juniper branches being a favored medium. Christianity, particularly the Roman Catholic, Coptic, and Greek Orthodox Churches, have retained the fumigant as a ceremonial purifier. In modern times, when one is taken ill by an inexplicable cause, it is not unknown for Greek Orthodox priests to make a house call to the afflicted, and use, among other things, fumigants to banish the illness-causing spirits or influences.

The magical virtues of fumigants were also well known to those promoting the witch-hunts of the Inquisition, and though associated with sorcery and diabolism, this did not prevent the good fathers from using them for Christian means. Johann Christoph Frölich, a professor of law in the late 17th century, advises that the torture chamber for extracting confession from witches be "constantly sprinkled with holy water and a smoke be made with blessed herbs." Valerio Polidori's book *Practica Exorcistarum ad Dæmones et Maleficia de Christi Fidelibus Pellendum* of 1626 gives three incenses (*Profumigatio Horribilis*) for driving away devils:

I. *6 drachms each of seed of Hypericum, Rue and Frankincense.*

II. *1 drachm each of Frankincense, Storax, Galbanum, Laudanum, and Garyophyllum.*

III. *1 drachm each of Galbanum Sulfur, Asafoetida, Aristolochia, and Rue.*

centration camps, US gas chambers for judicial executions, and also stockpiled as a chemical weapon by both the United States and the Soviet Union.

Agrippa relates, though in guarded language, that "the souls of the dead cannot be called up without blood and a carcass: but their shadows to be easily allured by the fumigations of these things..." In addition to propitiation and control of spirits, fumigants had other uses in sorcery, even to the point of having magical nomenclature. Divination by smoke, particularly using burning poppy seeds, was known as capnomancy; libanomancy was a more general term for divination by the fumes of incense.

The eleventh century manual of astrological sorcery *Picatrix* contains a series of sophisticated incense formulations, mainly for planetary genii, whose composition bears a close resemblance to suffumigants later appearing in European grimoires. The Incense of Saturn, is notable for its inclusion of opium, together with myrtle, saffron, plantain, bush grass, soiled wool, colocynth grease, black cat hair, and black goat wool.[9]

The publication of Henry Cornelius Agrippa's *Three Books of Occult Philosophy* in 1531 marks an important point in the development of the magical fumigant; this large tome was the largest collection of pagan neoplatonic magic ever compiled, and was of great influence in its time and for centuries to come. Two chapters are relevant to our study: "Chapter XLIII: Of Perfumes, or Suffumigations, Their Manner and Power" and "Chapter XLIV: The Composition of Some Fumes Appropriated to the Planets." A number of these recipes contain poisonous plants of known or suspect psychoactivity, for example:

> *So they say, that if of coriander, smallage, henbane and hemlock be made a fume, that spirits will presently come together; hence they are called the spirits' herbs.*

9 Also of note in the *Picatrix* are numerous other incenses using frankincense and saffron, as well as a suffumigant for unbinding magical talismans which utilizes the seed of mandrake.

Pertinent to our investigation of the fumigants of grimoires, however, is that locus where sorcerous control of spirit and psychoactivity intersect. A final recipe, perhaps more relevant to our investigation, is a Bavarian *Hexenrauch* or 'witches smoke' preserved in folklore, and used 'against evil', containing elderberry, henbane, vervain, fern, cannabis, belladonna, poppy seed, mint, rue, tansy, greater celandine, and maidenhair fern.[10]

XXII. *Magicians in circle conjuring a spirit, with horn brazier for suffumigation.* Das Kloster *grimoire corpus, circa 1845.*

We shall now examine some fumigant recipes as they appear in the grimoires and works of allied magic, with especial attention given those plants with known pharmacological activity. These recipes constitute only a portion of total incense recipes from the grimoiric corpus; there are many other formulae which feature plant ingredients not known to produce visionary effects or altered consciousness, whose provenance is chiefly of planetary correspondence, desirable odor, and that which some researchers call 'sympathetic magic.'

10 Fischer, L. "Ein 'Hexenrauch,'" *Bayerische Hefte für Volkskunde*, cited in Duerr, p. 142, n. 11.

Liber Juratus, or *The Sworn Book of Honorius*, is a grimoire dated to at least the 13th century which is particularly rich in fumigant recipes. The text informs the reader of "a suffumygacion made of these hearbes as foloweth causeth visions of the earthe to appeare", probably referring to the summoning of earth-elementals. The incense mixture is made from root of cane reed, root of giant fennel, pomegranate skin, henbane, red sandalwood, black poppy and "the hearbe tassi barbassi."[11] Another recipe is made from coriander, henbane, and skin of pomegranate and causes "visions in the air or elsewhere to appear."

The common psychoactive ingredient of known efficacy in these two incenses is black henbane (*Hyoscyamus niger*), a plant of the nightshade family long associated with diabolism. Together with wolfsbane (*Aconitum napellus*), poison hemlock (*Conium maculatum*), and a number of other nightshades, henbane comprises the essential botanical template for the *Unguentum Sabbati*, or witches' flying ointment, which are also contemporary to the time frame of our study. Henbane and other solanaceous plants bear a class of chemicals collectively referred to as the tropane alkaloids, the most important of which are atropine, hyoscyamine, and scopolamine. These alkaloids—particularly scopolamine—have long been known to inspire visions with strong phantasmagoric or erotic content; delirium, frenzy, and varying degrees of amnesia upon return to the senses.[12] The smoke of burning *Hyoscyamus* is also known to be psychoactive, with historically attested uses in Asia, the Near East, and North Africa. Of particular interest are is the usage of the smoldering leaves by Bed-

11 Most often considered to be mullein (*Verbascum thapsis*), known in folklore as Hag's Taper. In European and Asian magical traditions, mullein was held to possess the power to chase demons and noxious spirits. In some streams of lore, burning mullein stalks illuminated the night flights of witches. giant fennel may refer to apiaceous aromatics such as galbanum, armoniac, or asafoetida, or indeed poisonous plants in the umbel family such as hemlock.

12 Richard Evans Schultes, *Plants of the Gods*, p. 86; Hansen, *The Witches' Garden*, pp. 63–67.

ouin sorcerers and thieves for its stupefying effects.[13] Ott cites the experiments of Gustav Schenk (1954)[14] inhaling the vapors of burning henbane seeds, the psychoactive efficacy of which I have been able to verify personally; clearly the use of henbane in the form of a smoldering fume, especially employed indoors with no ventilation, and in conjunction with other ecstatic technologies, could give rise to visionary and ecstatic experience.

Poison hemlock is frequently found in ceremonial suffumigant recipes. Agrippa states that an incense made from sagapen, juice of hemlock, henbane, mullein, red sandalwood, and poppy, makes "spirits and strange shapes appear" as well as a second recipe containing hemlock but of simpler composition. Agrippa's demon-summoning formula is echoed by Girolamo Cardano, in his 1550 encyclopedia of natural science entitled *De Subtilitate*, who notes that vapors of smoldering coriander, celery, henbane, and poison hemlock will "instantly cause demons to assemble."

A major toxic principle in poison hemlock is the alkaloid coniine, considered a carcinogen, occurring in all parts of the plant, but concentrated strongest in the unripe fruits. Symptoms of poisoning include retching, vomiting, increasing muscular weakness and pain, rigor of the limbs, inability to speak, blindness, respiratory dysfunction, and death. Coniine paralyzes motor nerves, ultimately causing death by respiratory failure. Toxicological studies on coniine poisoning from oral ingestion indicate little in the way of visual distortion or hallucinogenic activity; and in fact one symptom is blindness.[15] Routes of coniine absorption other than oral are a somewhat different matter. In the context of discussing the Witches' Flying Ointments, Hansen cites unidentified clinical studies which indicate coniine ingest-

13 Penacchio et al, pp. 102–103.

14 *Pharmacotheon*, p. 369.

15 Turkington, Carol. *Poisons and Antidotes*, p. 132.

ed orally in small doses or rubbed into the skin can produce the sensation of gliding through the air.[16]

My own experience of accidental transdermal poisoning, in contexts unrelated to magical incense, confirms this to some degree. The dominant sensation was acute physical constriction, combined with lightheadedness, vertigo, and, in seeming, an increased sensitivity to light. Removed to an area of darkness and quietude, a sensation of lateral undulating floatation was noted, several inches above the ground attended. The sensation was in no way ecstatic, and indeed was pervaded by a constant and tenebrous fear of death.

If transdermal absorption of fresh plant alkaloids may be considered 'entheogenic', burning the plant is another matter. Coniine is an unstable alkaloid and begins to break down soon after the plant is harvested. Dried leaves and roots contain almost no coniine. The alkaloids are some- what more stable in the seed, and these have been harvested for incense usage in magical circles in which I have participated. In this particular case it should be noted that the suffumigants were initially not meant to be consciousness altering; rather, seed of Hemlock was incorporated for their planetary resonance, usually that of Saturn. Mixed with other ingredients of sweeter and more bouyant character, hemlock did not dominate the mixture's aromatics or energetics.

Smoldered alone, the seeds of *C. maculatum* are known to generate headaches and brief disorientation, but nothing close to atmospheres conducive to visionary trance or ecstatic intoxication. A 1998 report from my private record of magical operations is instructive. Here, four braziers were set in the inner chamber of a cave, each filled with an abundance of burning hemlock seeds:

16 Hansen, p. 74.

> *The smoke was opaque, plentiful, and at first unremarkable, having a carbonaceous odor and in no way intrusive to the operation. Within minutes, however, it grew acrid and became a stench, like burning tar or hair. As the smoke thickened, the lamps in the cave grew dim...it had a putrid, oleaginous quality, like a vaporized oil whose minute droplets fiercely adhered to the nasal passages and throat. At the cellular level I seemed to be experiencing a mounting somatic outrage or revolt; seeing was made difficult, breathing impossible. In less than ten minutes we were driven from the cave and spent the better part of an hour fighting off a smoke-induced migraine. Due to the foul nature of this incense, presumably due to volatilized compounds, much of the regalia was ruined, especially that made of cloth and paper.*[17]

When fresh, hemlock is an odd-smelling plant, often described as foul and compared to rodent urine; the odor of the fuming seeds is kindred, but marked with a sharp, invasive tinge of empyreuma. If a demon can be described in terms of being a singularly wretched vapor, smoking Hemlock will duly summon him, for it exceeds in foulness both sulfur and brimstone. Still, it must be noted that the poisonous denizens of the Apiaceae or Dill Family are numerous, with many exemplars bearing folk names 'hemlock' and being as toxic or more so than *C. maculatum*. In addition, several of the historical grimoire incense recipes explicitly mention the juice of the plant, not the seed.

Opium gum, or products of the opium poppy (*Papaver somniferum*) also occur with some frequency in grimoiric suffumigants. We have already examined two incense recipes, one from the Greek Magical Papyri using pium; the other from *Liber Juratus* for summoning earth-spirits, using black poppy in combination with Henbane. A second fumigant from *Liber Juratus* is compounded from "coriander, and saffron, henbanne, parslie

17 *Hypnotikon*.

and blake popie the water therof the popie dysstilled and tempered wt the iuce of the poungarnet skine." It is unclear what part of the poppy is first mentioned, but the "water of the poppy distilled" is probably tincture of opium or laudanum, or some other fluidic concentration of the plant's narcotic virtue. The same grimoire gives a suffumigant recipe for calling forth the planetary genius of Saturn, compounding seeds of henbane and poppy, together with root of mandrake, myrrh, ground lapis lazuli, and the blood or brains of a bat. In this recipe, smoke from the seeds of black poppy is probably insufficient to account for an appreciable narcotic effect when burned, but the addition of henbane and mandrake would certainly lend pharmacological potency to the mixture, and, unlike coniine, tropane alkaloids are notoriously stable over time and a wide platform of administration routes.

We now turn our attention to *The Sword of Moses*, a 10th century Hebrew-Aramaic book of magic in which we find the following spell:

> *To walk in the street and not to be recognized by anyone, take wormwood, perfumes, and soot, and moke thyself with it, and take the heart of a fox, and say the 'Sword,' and go out in the street.*

Though this fumigant is clearly burned for invisibility, it is notable for the presence of wormwood (*Artemisia absinthium*, an asteraceous herb best known as the 'Green Fairy' whose distilled spirits engendered the popular inebriant Absinthe. Wormwood contains Thujone, an isomer of camphor, with known psychoactivity. A second Wormwood-bearing fumigant recipe appears in the *Clavicula Solomonis*: a suffumigation for the planetary genii of Mars, consisting entirely of desiccated wormwood and rue. Wormwood, along with such herbs as poison hemlock and monkshood, in classical times fell under the dominion of Hekate, Goddess of sorcery.[18] Its use as a suffumigant is also known in modern collations of necromantic spells.[19]

Thujone in Wormwood is chemically related to camphor, a crude resin extracted from the lauraceous camphor tree (*Cinnamomum camphora*). It is also naturally present in other plants such as rosemary and mint. This refined botanical substance has been banned or regulated in some countries, including the United States, and its potency as a toxin should not be minimized. In addition to less pleasant side-effects of camphor poisoning, "dizziness, disturbed vision, delirium, and convulsions" are noted.[20] Its action upon the heart in certain cases may also classify it as an exhilarant. The resin appears in several grimoiric suffumigants of note, and is of known psychoactivity: Christian Rätsch observes that camphor "in high doses can elicit powerful states of ecstatic inebriation."[21]

Turning to our corpus of 'black books' we find a number of significant camphoraceous incense recipes. Echoing Agrippa's recipe for a Lunar suffumigation, *Liber Juratus* employs an almost identical recipe employing camphor and seeds of white poppy.[22] Camphor's chemical relationship with Thujone is significant

18 Agrippa, paraphrasing Porphyry, writes: "Hecate commanded how images should be constituted to her, and that they were to be surrounded with wormwood, and that domestic mice were to be painted, and the finest ornaments such as were most pleasing to her, and so many mice as her forms were to be taken; then blood, myrrh, storax, and other things were to be burnt: which things if they were done, she would appear, and answer the worker thereof by dreams."

19 Wright, Elbee. *Book of Legendary Spells*, p. 40. This use of Wormwood is often quoted as a medieval spell in modern compilations of magic and herb-lore, such as *Herbs in Magic and Alchemy* (C. L. Zalewski).

20 Blumgarten, A. S. *Textbook of Materia Medica*, Pharmacology and Therapeutics, 7th ed. MacMilllan, New York, NY 1942.

21 *A Dictionary of Sacred and Magical Plants*, p. 68.

22 Kenneth Grant notes the magical relationship of Camphor to the Lunar current, the subtle emanant qualities of the sexual elixirs, and also to Śiva and his retinue of powers. See *Aleister Crowley and the Hidden God*, p. 116.

and forms a link in incenses with Wormwood, which has shown mild psychoactivity when burned and inhaled.

In the *Grand Grimoire*, the conjuration of the devil Lucifuge, Prince and Master of Rebel Spirits, is prefaced by burning copious amounts of pure camphor in the brazier. Camphor-bearing incenses are found in other grimoires: in the section concerning suffumigations for invoking planetary genii from *The Key of Solomon the King*, we find a Lunar fumigant compounded from white sandalwood, camphor, aloes, amber, pounded seeds of cucumber, artemisia, and ranunculus. This assembly includes wormwood and ranunculus, a plant of significant toxicological standing.[23]

Helleborus niger or Black Hellebore, is a poison of decided infamy; all parts of the plant render up potent venoms. The extract of the rhizome was used in antiquity and medieval times as an utensil of murder, most notably by King Attalus III of Pergamos. The glycosides heleborin and hellebrin also assert a strong presence in the plant; chemically related to telo-cinobufagin (a venom of toad skin), they affect cardiac muscle in a similar manner as foxglove, slowing heart rate. At least two grimoire incenses employ black hellebore. One is listed among the suffumigations for planetary genii from *The Key of Solomon the King*:

> *Spell for vivifying a talisman with the genius of Saturn: 'This consecration consists in exposing the talisman to the fumes of a scent composed of alum, assa-foetida, scammony and sulphur,*

23 The genus *Ranunculus*, or true buttercups, comprises some 400 species, many of which contain important poisons (See Stary, pp. 166–171). Perhaps the most important are Meadow Buttercup (*R. acris*), Lesser Celandine (*R. ficaria*), Creeping Buttercup (*R. repens*) and Cursed Buttercup (*R. sceleratus*). The chief toxins are ranunculin and protoanemonine; these manifest poisoning symptoms, sometimes fatal, when ingested. Many species of Ranunculus are irritant vesicants and raise blisters when applied topically. Data on the phytotoxicity and psychoactivity of buttercups when burnt and inhaled is apparently lacking.

> *which are burned with cyprus [sic.] ash, and stalks of black hellebore lighted in an earthenware chafing-dish which has never been used for any other purpose and which has to be ground to dust and buried secretly in an unfrequented spot after the operation.'*

For conjuring the planetary genius of Mars, the *Liber Juratus* recommends a fumigant made from euphorbium, bdellium, armoniac, two kinds of hellebore, myrrh, brains of a raven, human blood, blood of a black cat, sulfur, and "the powder of the stone called magnes." Severe Hellebore poisoning is a grim affair: before the bleak twilight of unconsciousness or cardiac arrest, the victim suffers burning of the mouth, slavering, a great roaring in the ears, and visual aberrations. Some persons report visionary effects and delirium from smoking the dried leaves[24] but given the disagreeable relationship between Hellebore toxins and the human organism, this is discouraged.

A number of other plants of known psychoactivity appear in other fumigants used as adjuncts to grimoire incenses or in later magical usage. One medieval rite of thaumaturgy made use of burning cannabis as a fumigant while invoking spirits for the purpose of a love spell.[25] Another incense is documented as being employed in the 1700s by le Chevaliers Elus Coens, the magical order of Martinez de Pasqually:

> *The most important ceremonies took place during the first quarter of the moon and were accompanied by the use of an incense compounded of storax, olibanum, saffron, poppy seeds, agaric spores, cinnamon, nutmeg and mastic; at least two, and*

24 An infrequent but known practice in certain rites of modern Essex witchcraft; I thank Frater A. for permission of reference here.

25 Wright, pp. 69–70. This action followed casting of cannabis seed powder to the Four Winds, beginning in the East, and moving in a counterclockwise direction.

> *probably more, of these substances are possessed of hallucinogenic properties, so it is not surprising that the Elect Cohens seem to found their magic effective.*[26]

While this incense contains poppy, which we have already examined, it is most interesting for its inclusion of 'agaric', probably the fly agaric mushroom (*Amanita muscaria*), a well known but highly variable entheogen. While it is doubtful that burning spores alone would suffice to provoke a visionary trance, there is some evidence, cited by Christian Rätsch, suggesting the smoke of the mushroom may be inhaled for inebriating virtues, particularly among the Chuj Maya, who smoke the dried mushroom with tobacco to induce a clairvoyant trance for healing. Saffron and frankincense, both possessing subtle effects alone, and possibly potentiated together, are also important. Nutmeg additionally possesses a complex of psychoactive principles which manifest euphoria and mild hallucinogenic activity when ingested orally, though whether these would become bioavailable when small amounts were volatilized in burning incense is unknown. It must also be remembered that synergistic effects with multiple botanical compounds are also known.

In *The Testament of Solomon* a smudge preparation is revealed to the wizard in the course of conversing with a demon:

> *But I Solomon questioned him, saying: 'If thou wouldst gain a respite, discourse to me about the things in heaven.' And Beelzeboul said: 'Hear, O king, if thou burn gum, and incense, and bulb of the sea, with nard and saffron, and light seven lamps... thou wilt firmly fix thy house. And if, being pure thou light them at dawn in the sun alight, then wilt thou see the heavenly*

26 Francis King, *The Secret Rituals of the* O.T.O., editor's introduction. Pasqually's magical system combined elements of Gnostic, Manichaean and Cathar currents, as well as Hermetica. Spirit-summoning and conversations with angels were also important.

> *dragons, how they wind themselves along and drag the chariot of the sun.'*

Aside from the fantastic serpentine visions described to manifest from the enchanted smoke, this formula is important due to its inclusion of saffron (*Crocus sativus*), a plant appearing in many ancient incense recipes, and also present in many of the aforegoing formulae. *Crocus* is the genus to which true saffron belongs, but in ancient recipes 'crocus' or 'saffron' often referred to a number of different iridaceous plants, some of which were many times more toxic, such as meadow saffron (*Colchium autumnale*). An incense found in *The Leyden Papyrus* to summon or reveal a thief, employed powdered crocus and Alum. An obscure fifteenth-century grimoire entitled *The Book of Angels, Rings, Characters, and Images of the Planets* employs a saffron fumigant for conjuring solar genii; it appears in another recipe in *Liber Juratus*, this one used to "cause visions in the air and the shadows of sepulchres of the earth to appear."

> *Take the naturall seed of the fyshe called a whalle, lingnum aloes, costus, muske saffronne, armoniacum, wt the blude of the foule called a lapwinge, and make a conffeccion therof, wt this sayde conffeccion make a fumigaccion in a conuenient place, and you shall see visyons in the ayer, take of the sayd conffeccion and make a fumygacion aboute the sepulkers and vissions of the dedd shall and wyll appeare...*[27]

Agrippa also lists the most powerful fumigant "according to Hermes", which contains one herb for each of the planets, including saffron to resonate the powers of Venus.

Though its aromatic and pigmenting virtues are known, the psychoactivity of *C. sativus*, when burned, is in debate. However

27 Fanger, Claire ed. *Conjuring Spirits: Texts and Traditions of Medieval Ritual Magic*, pp. 67–69. The manuscript in question is Cambridge Univ. Lib. MS Dd.xi-45.

when ingested orally it is considered aphrodisiac, and in excess narcotic. Saffron has also been found in at least one recipe for Witches' Flying Ointment.[28]

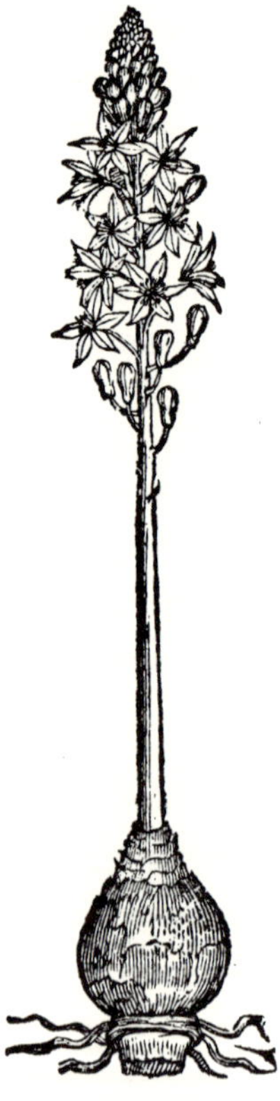

XXIII. *Maritime Squill* (Drimia maritima).

Another ingredient of interest in the *Testament of Solomon* incense is 'Bulb of the Sea.' While there is some speculation about precisely what plant the demon Beelzeboul is referring to in the text, it is likely squill (*Scilla* spp.; *Urginea maritima*), a hardy plant of the Lily Family widespread in the Mediterranean and having a large, enspissated bulb. The vulgar name by which this plant was most commonly known is Sea Onion. Though traditional use of squill as a mind-altering plant is not documented, the plant does have a rich magical pedigree for hex-breaking and purification. The bulb of Maritime Squill (*Drimia maritima*) was, and still is, affixed to doors as a magical specific against unwelcome visitors and noxious spirits. The ancient Greek ceremony of the *pharmakos*, or Scapegoat, involved taking the human victim to the place of sacrifice and fed with figs, cheese, and barley cake; flogged seven times on the genitals with squill; and then burnt on wood gathered from wild trees. The ashes of the victim were later scattered to the winds of the sea. Additionally, the pharmacological profile of

28 Hansen, p. 90. The quote is from a 1626 publication by Sir Francis Bacon: "The ointment that witches use, is reported to be made out of the fat of children digged out of their graves; of the juices of smallage, wolf-bane, and cinquefoil, mingled with the meal of fine wheat. But I suppose, the soporific medicines are likeliest to do it; which are henbane, hemlock, mandrake, moonshade, tobacco, opium, saffron, poplar leaves, etc."

Sea Onion is interesting. *Urginea maritima* contains such cardiac glycosides as scillaren A and B and scillirosid. Mild poisoning results in inflammation and cramping in the gastrointestinal tract; distortions of the vision may also occur, as well as heart irregularity. Though what we know about Squill poisoning usually comes from overdoses when administered orally as a medication, it is plausible that the burning smoke of this plant might contain the same or related poisons. Its linkage with spirit-summoning incenses may also stem from the Levantine folklore associating the plant with graveyards and the spirits of the dead.

In considering incense adjuncts of animal origin, mention must also be made of Ambergris, an aromatic gastronintestinal secretion of the sperm whale used in the perfume industry, which also appears in grimoric suffumigation formulae. In magic and folk medicine Ambergris is traditionally considered an aphrodisiac, added to food or tonics and worn as a fragrance. Assays of the substance reveal that it contains substantive amounts of cholestanol sterols and the fragrant compound ambrein, recently shown in animal trials to increase arousal and sexual behavior.[29] Given the adrenergic effects of sterols, as well as hormonal roles in the physiology of arousal, the volatilization of ambergris via incense likely plays a significant role in the somatic and psychic states during and immediately following suffumigation.

The phenomenology of spirit-conjuring incenses raises a multitude of questions and possible approaches for the practitioner. Certainly, a detailed pharmacopoeia of venomous incense plants is a necessary component for the study and mastery of this Art. Chemically, the grimoric incense formulae are complex witches' stews with many botanical ingredients. However, it is necessary to recognize that a great many vectors align to promote their efficacy of fostering a 'bridge of smoke' to the

29 Taha, S. A. et al. "Effect of ambrein, a major constituent of ambergris, on masculine sexual behavior in rats." *Archives Internationales de Pharmacodynamie et de Therapie*, 1995 Mar–Apr; 329(2):283–94.

spirit-world. Together with our suffumigation *materia magica*, all, when approached wisely, provide a deeper understanding of the censing Art and an elementary patterning for its magical refinement.

The rational toxicologist will, according to inclination, focus wholly upon the biochemical constituents of the suffumigation and their physiological effects. Here, the essential question becomes: does the smoke of the burning or smoldering incense ingredients elicit a somatic response consistent with the appearance of demons, angels, gods, spirits of the undead, and their oracular transmissions? Where affirmative, the underlying botanical skeleton of such incenses in the first instance is comprised of ingredients whose psychoactive poisons are active or potentiated when volatilized through the agency of fire or heating. Opium, henbane, camphor, wormwood, hashish and other cannabis products, all of which are found within grimoiric incenses, are well situated within this category. A second tier of adjuncts includes fragrant arboreal resins such as frankincense or juniper, with lesser but established psychoactivity, to assist the trance state and provide aromatic counterbalance to the often foul odor of burning carbonaceous plant materials. Non-psychoactive ingredients include pure aromatics; binders such as gum arabic or the organs of certain animals; coloring agents such as saffron; incendiary or accelerant adjuncts such as charcoal, sulfur, or saltpeter; and smoke-producing agents. Ingredients which are added for their talismanic qualities are also an important category, and there are many ingredients which possess more than one of these properties. Beyond the individual contributions of each suffumigant ingredient, we may also posit unique synergistic interaction between them which gives rise to novel psychophysiological effects.

In addition to volatilized psychoactive compounds and aromatics arising from magical incense, the presence of carbon dioxide and monoxide—both toxic and capable of producing dizziness and disorientation—are present in the ritual operation during suffumigation. Carbon dioxide, a lethal poison to animal life in surprisingly low proportions to Oxygen, has also been used as a therapeutic drug, anaesthetic, and as a psychoactive to deliberately alter consciousness. The hypoxia induced by autoerotic asphyxiation, when it does not result in death, is one such use, attributable to mounting levels of CO_2 in the blood and the endorphin release associated with orgasm. States of terror and panic, also common during the conjuration of certain spirits, are part of an initial physiological reflex to carbon dioxide intoxication, as Pendell notes.[30] The prophecy-inducing vapors of the Oracle of Apollo at Delphi, described as emanating from a telluric chasm during certain periods of its ancient use, have recently been proposed by science as a mixture of CO_2 and H_2S, periodically released during fault rupture.[31] Characteristic of carbon dioxide intoxication is initial fear and feelings of suffocation followed by peaceful sensations and synaesthesia, appropriately situated at the veil of death.

In assessing the efficacy of spirit-summoning incenses, consideration must also be given to the locus of the magical operation. Certain grimoire images illustrate operations of magical conjuration situated out of doors, in graveyards, ruins, forests, or remote places. But such manuals of sorcery were disseminated through a kind of magical underground, and spirit-summoning operations during the age of grimoires were often illegal and carried real consequences. It is therefore reasonable to assume that a great many of these rites were carried out inside buildings, churches, or enclosed crypts, where the incense fumes would rapidly increase and concentrate, rather than be dispersed in the open air. The same principle is operative in the Native American

30 *Pharmako/Gnosis*, "On Elementals", pp. 265–274.

31 Piccardi, Luigi, et al. "Scent of a myth: tectonics, geochemistry and geomythology at Delphi." Journal of the Geological Society 2008; v. 165; pp. 5–18. The scholars highlight the link between the site's original name *Pytho*='rotten' and the stench of Hydrogen sulfide fumes. The site became sacred to Apollo after he slew its serpent guardian; the mantic fumes arose from her corpse.

purification rite of the sweat lodge, and has led to unintended fatalities when the rite has been appropriated outside traditional culture.

An historical imagining of grimoric spirit conjurations must also consider contemporary understanding of spirit-natures and their relation to the influence of air, odors, and vapors. Written spirit-conjuration rites, as a magical genre, have in varied eras portrayed the entities as composed of fire and smoke, of amorphous or partially congealed shape, a face or torso suddenly manifesting from a churning brume. In other instances, conjured spirits are described as the smoke or fire itself, a morphology similar to that of the Hebrew god Yahweh, variously portrayed as formed of smoke, fire, or cloud.[32] This fulminant guise, in terms of the appearance of the divine spirit, is similar in many respects to the *jinn* of Arabian and Islamic lore, considered angelic or diabolic beings depending on spirit and context. The nexus of this magical lore with an area possessing an ancient history of fire offerings, and advanced traditions of incense formulae should not surprise us.

More relevant to the operations of grimoiric conjuring is John Dee's *Compendium Heptarchiæ Mysticæ* of 1588, a record of his angelic spirit summoning. The work contains physical descriptions of some of the entities summoned, several of which appear as amorphous wisps of vapor. One of these is Princeps Bralges, who "to my seer appeared like little smokes, with out any forme"; another group of spirits were the 42 ministers of the spirit Bynepor, who were "like ghostes or smokes without all forme having every one of them a little glittering spark of fyre in the middest of them: and every spark a letter in it." In ecstatic rites the human soul itself is frequently beheld as a vapor, as with the egress of the souls of the night-walking *benandanti* in the form of smoke, and

32 *Book of Exodus*, passim.

the well-known passages of witches up chimneys and through locks.[33]

The votive, offertory aspect of the suffumigant as food for the spirits, or a vehicle for conveying the *numen* of prayer and adoration must also be considered as a portion of the art. The notion that censing perfumes are irresistible to spirituous entities permeates both religious and magical usages of incense. In the European grimoire corpus, this phenomenon is perhaps best illustrated in the grimoire *Ars Almadel*. Using a suffumigation of mastic (*Pistchia lentiscus*) the scribe remarks:

> *as soon as the angell smells it he beginneth to speake with a low voice asking what your desire is and why you have called the princes and governors of his Altitude...*

The documented appearance of spirits assuming the forms of vapor or smoke is balanced by other grimoires containing fanciful descriptions of the spirits invoked. However, as is often the case in the magical art of scrying, one value of the vapor cloud lies in its simultaneous opacity, monochromatic hue, and fluidity of form, which under precise ritual conditions could congeal or harden into a fixed form before the eyes of the operator. This admixture of physical virtues is essential for the ancient magical disciplines of capnomancy and libanomancy (divination by smoke) and aeromancy (divination by clouds).[34] The inquisitor Pierre deLancre reported that one method by which the Basque witches could travel to the Sabbat was to *emit through their mouths a certain thick steam in which they see all that happens there, as if they are seeing in a mirror.*[35]

33 Ginzburg, *Ecstasies*, pp. 166.

34 Agrippa: "The like things reports Dion in his Roman History, in a place which they call the Nymphs: where frankincense being cast into the flames, oracles were revived concerning all those things which he did desire to know, especially concerning death, and those things which belonged to marriages."

35 De Lancre, *On the Inconstancy of Witches*, p. 113.

In medieval magic, air, and in particular vapors and odors, were acknowledged carriers of power, for good or ill. Agrippa was precise on the matter of good and bad odors and their magical jurisdiction. Deferring to the "opinion of magicians", he states:

> *In every good matter, as love, goodwill, and the like, there must be a good fume, odoriferous and precious; and in every evil matter, as hatred, anger, misery, and the like, there must be a stinking fume, that is of no worth.*

The early corpus of Christian demonological literature played also some part in this: the early church father Justin Martyr, relates that demons oppress mankind *partly with magic, partly with fear and punishment, partly with the discipline of sacred odors burnt.*[36] Evil vapors, as well as good, could be manufactured as well as naturally occurring. Felix Hammerlin (1389–1460), a church authority at the Council of Brasle, wrote in his *Dialogus de Nobilitate et Rusticitate*, that women "boil in a pot foul and venomous herbs and things, and then upon exposing it to the rays of the sun a vapour arises to the second region of the air and condenses into clouds which send down hail and tempests." Though such descriptions may or may not be viewed as "witch-hunt propaganda," it clearly reveals a complex cartography of vapor belief, as well as notion of a prototypal grimoire fumigant.

Considering these perceptions of the appearance of spirits we see a cosmology in which air is not only responsible for the transmission of evil and good, but also for the generation of demons and angels: it is the very stuff whereof they are composed. Thus is conveyed the notion that the powers of these entities, as manifest through air, are inherent in both material and action, force and form. Angelic and demonic entities were also thought to be formed of fire. Air and fire are both necessary in the action of suffumigation: fire to ignite and transform the perfume,

36 *Apologia ad Antoninum Pium*, c. AD 165.

air to distribute the smoke and nourish the flame. An historical user of grimoire magic whose cosmology included these concepts would probably regard a suffumigant as a material base allowing greater ease for the manifestation of a spirit. This belief combined with prolonged chanting, praying, and circumambulation, could easily potentiate an incense comprised of psychoactive plants and contribute to the sorcerous experience of congress with spirits.

Standing within the fuming vortex of the magic circle itself, it quickly becomes clear to the operator that all diverse factors of suffumigation must converge in alignment for the efficacy of the operation. Prolonged incantation of sacred or barbarous words, the obsessive focus upon singular telesma or sigils, the nexus of Will, Desire and Belief, and the restriction of certain senses and the elevation of others, are of themselves often sufficient to generate trance states. The perfumed aromatic, accessing a favorable stratum of magical correspondence, builds and multiplies a refractory ambience which is an extension of the enchanted circle itself. The atmospheres thus generated are then become receptive to a further key: chemical psychoactivity. In greater or lesser amounts, this element may serve to throw wide the gates where other factors have merely cracked them, and as an additional armament of the magician of the Poison Path.

EDEN'S BODY

One meets, along the path, the sorcerer who wears no clothes, reads no book, and fashions no implements or tools. In the places of desolation, he or she calls upon the gods with the living blood and bones, and the somatic permutations that cast the circle as the Living Round of Flesh. This stance is in direct contrast to a strongly materialist occultism: its audacity often offends the book-learned, but the force of its appearance, conviction and efficacy remains. When such adepts are asked about this approach, one frequent response is that the Great Temple lies within, and that all external objects are mere encumbrances or distractions. For some it is an act of humility before the spirit-world; for others a transgressive state, overturning societal norms of modesty and property ownership. Whether it be the night-frolicking revelers of the medieval Witches' Sabbath, the ash-smeared Aghori, or the Transylvanian maidens dancing nude before the *Mandragora*, the unadorned sorcerer is a potent image of power and communion with the spirits.

In obscure magical tributaries of witchcraft praxis, this concept is known as the *Edenic Body*, the concept of 'the Garden Within'—a state of primordial power, its bowers bearing the strangest and most potent of fruits—before the imposition of exterior 'garb.' It posits the body as an arena of sorcerous power, a series of magically-attuned instruments and zones of numen capable of generating ecstasy, receiving occult knowledge, and performing miracles. In such a conception the magical radiance of the Body is sovereign, and self-contained, but it is also permeable and penetrable by the *Other*, existing as a potential for infinite magical power. Above all, it expresses the Arcanum that, in order to manifest the work of the Art Magical, power must be embodied.[1]

The obsession with metaphysical Eden is not confined to Jewish and Christian theologians; it has been the preoccupation of occult and heretical groups since the story was recorded in writing, and likely earlier. Metaphysical extrapolations on the Green Primordium abound in such Gnostic scriptures as *Apocalypse of Adam*, and non-canonical gospels such as *The Life of Adam and Eve*. Similar magical books of alleged Edenic origin are extant, such as the medieval kabbalistic text *Sefer Raziel HaMalakh*, said to have been revealed to Adam by the angel Raziel. John Dee, one of the most important figures in western magical traditions, during his spirit-discourse with the angel Uriel, is taught the Edenic and pre-Adamitic origin of a magical book in his possession, the *Book of Soyga*.[2]

In the teachings of Sabbatic Witchcraft, the Edenic concept of the First Body is closely linked with the metaphysics of poison, chiefly as a dispensation of the rebel angel Samael. However, instead of focusing upon this transmission as the source of the 'Original Sin' or 'Fall' of mankind, it is seen as the primordial Initiation, an Ascent of the status of humanity or 'Raising' to an

1 The Arcanum of the Revealed Garden, being the occult powers of herbs in witchcraft, was the subject of my book *Viridarium Umbris: The Pleasure-Garden of Shadow* (Xoanon, 2005). The Arcanum of the Concealed Garden is the subject of a separate volume.

2 *Liber Primus*, March 10, 1582.

angelic or deific state. Such a state is implied in the orthodox concept of 'Knowledge of Good and Evil', but within occult traditions, 'knowledge' is often a cipher for hidden power, conveyed of spirit to an elite body of initiates. In Traditional Witchcraft, one may interpret such Knowledge as the Power of Healing, and the Power of Harming, an ethos which has been called 'Crooked Path Sorcery' because of its embrace of both zones. Additionally, the estate of 'opening the eyes' of Adam and Eve is considered cognate with the attenuation of bodily centers to newly-ingressed witch- power. The hypostasis of Adam in this all-realised state is known in the Sabbatic Tradition as *Malachadamas*, 'Kingly Man', also called the 'Trimorphic Attainer' who commands his uncreated, present-plenary, and all-potential states.[3] The station bears certain similarities to the kabbalistic concept of Adam Kadmon or 'Original Man.'

This transmission of power originates with Samael, the transgressive Serpent-Angel and tempter offering poison unto First Woman. Indeed, the Hebrew name Samael may be translated 'poison of God' or 'venom of God.' The name is also ascribed a Syrian origin in *Shemal* (meaning 'left'), the God of the Sabaeans sometimes identified as the prince of demons or djinn. In some ancient esoteric doctrines this angel incepted the growth of the very tree of Eden's heart, for he 'Planted the Vine which caused Adam to stray.'[4] In other kabbalistic lore, Samael forms a conjugal union with Lilith, also a patroness of witches in some traditions, and further linking him with a bodily—in this case sexual—transmission of power.[5] The Serpent-Angel remains, in some angelic hierarchies, a servant of the High God; although

3 By gematric permutation, Malachadamas=484, or 222 being the number of copulative permutations of the 22 Sacred Letters of the Witches Alphabet. See Chumbley, *Azoëtia* (1992, 2002), and Schulke, *Viridarium Umbris* (2005). The arcana of Malachadamas is treated in Chumbley's unpublished work *Auraeon* (2004).

4 III Baruch 4:8.

5 *Treatise on the Left Emanation.*

he acts to seduce humankind or lead them astray, the implication being that poison and deceit, if not holy in themselves, at least operate under divine authority. Whether the Fruit he proffered was itself poison, or whether it became infected, the scriptures do not agree upon; in some legends Samael 'injected' his venom into the fruit of the Tree.

The transmission of poison, first from angel unto tree, thence unto Eve, and on from Eve unto Adam, conceals an initiatic formula of the Poison Path: the principle of the *transmigration of venom*, which accounts for the transfer and evolution of destructive principles from one body to another. Ibn Wahshiya's *Book on Poisons* describes the production of a lethal compound beginning with the confinement of a mouse and scorpion inside a brazen vessel. After the creatures expire from mutual attack, the corpse of the rodent serves as the base matter of the developing venom, fermented in a leaden vessel with spurge,[6] opopanax, mustard, aloe and saffron. This principle also participates as an evolutionary strategy in the botanical world, as in such plants as the semi-parasitic mistletoe, whose medicinal virtues result from the age of the plant as well as the peculiar phytochemical symbiosis shared with its host. Depending on the nature of its therapeutic use, the most potent mistletoe may be found growing on rowan, apple, oak, hawthorn, or, indeed, and of a number of other sylvan hosts. Though semi-parasitic in nature, the mature mistletoe, when harvested for its magico-pharmacological usage, is as much a radiance of its wedded host as any singular virtue on its own.

Similarly, certain poisons, when distilled through the Alembic of Man and Woman, may be potentiated in the humors of the body, and magically expressed therein. These operations may take place on the gross chemical level, as with the distillation of the fly agaric or psilocybin mushrooms, whose visionary compounds pass into the urine, largely scrubbed of other

6 Genus *Euphorbia*, known for its fœtid, irritant latex-like sap.

poisons by the liver. Under the precise physiological parameters of magico-sexual operations, this potentiation may also occur with astral poisons and nectars, which form the principal elixirs of the Corporeal Laboratory. Knowledge of this permeability between the phantasmal order (or the imaginal mind) and the subtle alchemisms of the flesh has been the preserve of tantrics and certain occultists, but also, on vastly different ontological levels, the witch-current. The matter has engaged the fascination of a considerable number of writers, but the process only assumes a living quality when engaged with the highest levels of private devotion.

Zones of occult power residing in bodily centers have been mapped extensively in the esoteric matrices of mystical and magical systems, as with the Tantric chakras of Hinduism and Bön. More recently occult cartography has included the *points chauds* or 'hot points' of the Voudon Gnosis system of Michael Bertiaux, which importantly assign both power and sentience to these foci. Traditional Witchcraft, especially the forms giving rise to the Sabbatic Current, also contains a unique Cartography of the Flesh. Some of these bodily zones of witch-power possess hypercorporeal attributes which may be extended into the magical realm, one term for this being 'Assumption of the New Flesh'. Mysteries of these fleshly zones of power have been re-presented in Andrew Chumbley's *Azoëtia,* in particular the organ of the Eye.[7] Certain oral teachings of Traditional Witchcraft also appertain a kind of *Ars Memoria* correlating magical powers to objects such as stones, trees, animals, or physical loci. These imbedded teachings find precise locations in the physical body and may be physically manipulated to express their magical potentials.

An important stratum of witch-power animating the Edenic Body is the incarnative perpetuity of atavistic wisdom, the procession of spirit-knowledge comprising and informing the

7 Chumbley *Azoëtia*, pp. 194–195. Also Schulke, *Lux Hæresis* for ritual methodologies for magical expansion of the senses.

present body. These dark reverberations of power operate on the deepest ancestral levels, and, being pre-human, are devoid of human morality and attribute. Brought forth by specific forms of ritual trance-induction, as well as the usage of plant poisons such as *Atropa belladonna,* they assist in the resonance of specific divinatory vectors, especially the non-vocal sensorial. When manifest in the body, they may assume a concrescence in the blood or sexual fluid, reminiscent of the mythic stone *Dracontias,* torn from the brains of living serpents and revered by the ancients. These 'philosophical stones' or material emanations, like the bezoars of the Alchemists, effect certain cures, such as healing poisonous wounds and served as protection from venomous animals. In spell-craft they assume the First Matter of the Living Fetish.

In the witch-cult, a traditional form of atavistic resonance lies in the magical embodiment of the Serpent-Angel Samael. Allied praxes venerate a preparation called 'The Serpent's Graal'—a draught prepared according to two essential formulations. The Lunar Cup, being the production of a sorcerous atavistic emanation, is vinted from the sexual emission of Eve and the Serpent, and prepared in accordance with the strictures of the Agapae.[8] The Solar cup makes use of *Mandragora,* the Mandrake or 'Man-Dragon' allied to the notion of the Embodied Serpent. Other formulations are for the Assumption of the Serpent Corpus are known; the Arcanum may manifest along differing trajectories but all are united in the Admonition of the Snake: *take, eat, and be wise.*

One of the primary routes of 'magical embodiment' is the use of the cultic fetish, the poppet or eidolon giving form to the spirit or *magistellus* of the witch. The European *alraun* traditions of sorcery using the carved root of mandrake (*Mandragora* spp.), a unique member of the nightshade grouping of plants sometimes called 'hexing herbs' is of especial relevance to the adept of

8 The Serpent in its aspect of the Kundalini force is also relevant to this version.

the Poison Path. At the center of the magical concerns of mandrake effigy-spells are enchantments for empowerment of the human body: increased fertility, sexual vigour, and attracting a sexual consort. The crudely anthropomorphic form of the root has been advanced as the reason for this, as many of the examples of mandrake effigies have pronounced phalli or vulvas.[9] However, certain preparations of the mandrake, especially in a fermented alcoholic form, behave as an aphrodisiac in small doses, and have had this reputation among cunning folk for centuries. More recently, scientific research has identified withanolides and sterols present in *Mandragora officinarum*, a species which has also amassed a considerable amount of aphrodisiac lore.[10] Curiously, many of the taboos surrounding the magical usage of mandrake fetishes demand a kind of marital, or consort relationship between the fetish and the operator, in which one pledges body and soul to the Root. Affirmation of the Pact through adoration of the fetish brings power, but neglect brings absolute ruin.

One of the largest and most diverse collections of mandrake fetishes presently resides in the Richel-Eldermans collection in the Museum of Witchcraft in Boscastle. Providing evidence for a continuation of archaic traditional mandrake sorcery into the twentieth century, the collection was, until recently, the privately-held magical cache of several obscure European magical orders. These carved figures are often accompanied by a specialized coffin for their interment and a number of the effigies appear to have been regularly anointed with votive offerings of blood, and likely sexual fluids as well. The effigies are accompanied by a considerable corpus of written and graphical material concerning Mandrake sorcery, much of it in magical cipher, and

9 In some exemplars I have seen in private collections, the entire root is ithyphallic, carved as an *olisbos* for ritual sexual penetration.

10 Suleiman, Rami K. et al. "New withanolides from *Mandragora officinarum*: first report of withanolides from the Genus *Mandragora*", 2010.

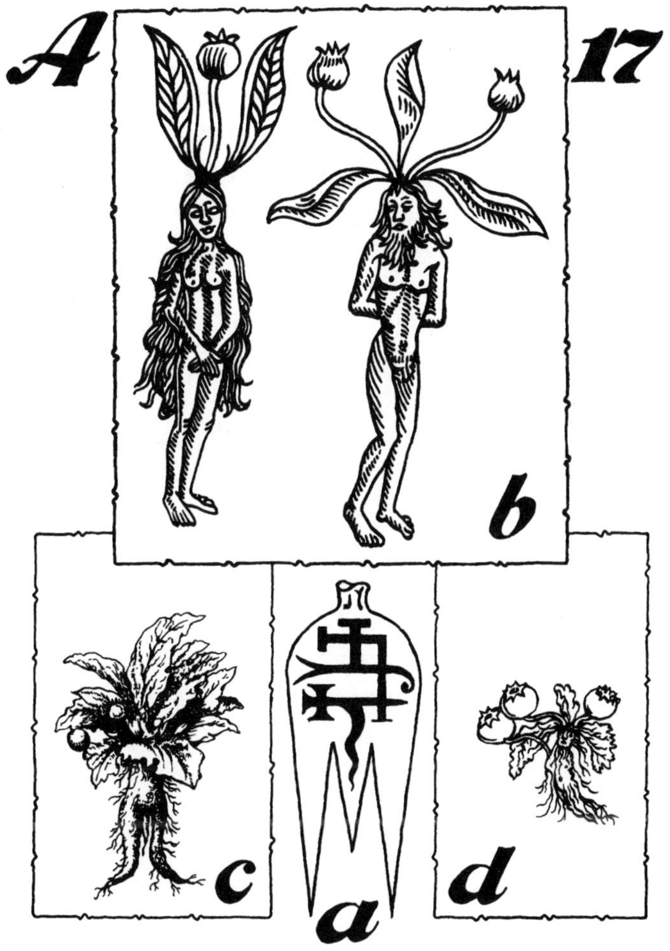

XXIV. *Womandrake and mandrake in cipher manuscript, from the Richel-Eldermans collection of artifacts, Museum of Witchcraft and Magic, Boscastle, Cornnwall.*

some highly sexualized. When considered in the context of the Richel-Eldermans collection as a whole, with its strong component of sexual sorcery, the function of the effigies, or their accompanying practices, were likely eroto-magical in nature. One image from the collection, showing male and female *Mandragora* plants, not only hypostasizes the plant into sexualized bodies but also suggests, by its symbolism, a connubium of the male and female essences, united by or within an object which appears to be a flask, cauldron, or alchemical vessel. The 'embodied magical' and 'sexual' arcanae of the mandrake are potentiated when one considers the lore of the plant's genesis in the ejaculate of the gallows-corpse, a mystery also represented graphically in the Richel collection. This conjures to mind the Latin phrase *Semen profusam Diabolo dicant,* 'dedicate the spilled semen to the Devil'—a motto of the witch-cult concealing a specific eucharistic formula of *magia sexualis.*

Further, my own personal work with the mandrake as a prepared ritual sacrament has yielded a strict protocol of preparation and administration which results in an ecstatic profile that is highly somatic. Skillfully prepared, its phases of action upon the bodily centers of power are at once aphrodisiac and illuminant, stimulant and hypnotic. This combination of mechanisms is difficult to attain with any single plant, and, for that matter, in the realm of synthetics or prescribed psychotropic medications. At once one receives and transmits power through the agency of the body, both modalities simultaneous to each other, and mutually pleasurable. However, this pathway of the sacrament is a narrow one, restricted by strict protocols of traditional preparation and administration, as well as accompanying magical technique. Falling short of the mark, the path may terminate in stagnant lassitude or else descend into a nightmare of sensorial abomination and narcosis.

The occult cartography of Eden's Body also posits that 'the internal garden' of the physical vessel may be magically tended to produce a harvest of endogenous chemical inebriants as the central work of the adept. These are attenuated and expressed through specific stimuli such as pain, orgasm, karezza, and strategies of physical exhilaration, exhaustion or quietus. Though properly situated in the scientific realm of Biochemistry, these 'corporeal potions' have occult applications, and are of specialized esoteric import to the Poison Path.

Sex-magical praxes for the expression of these corporeal potions, conveyed to a secret body of adepts, are known in many occult schools. The potency and magical potential of these biochemical elixirs is great, but the notion of locating a 'pantheon within' the flesh of the body has long been heretical. In the Edenic prohibition 'thou shalt not eat', some have glimpsed a cipher for veiling the mysteries from the eyes of the profane, whilst others regard it as the 'fall' of the demiurge himself, conquered by the insecurity of an initiated estate in his own creations. One often-made criticism of modern drug prohibition is that it is driven by fear of gnostic revelation or knowledge arising from bodily ecstasy, and the consequent shattering of state or religion-enforced paradigms. Though the phenomenon of drug prohibition transects cultures and is far too complex to be pinioned with a single motive, one must wonder if the same fears lie at the root of state repressions of sexual freedom and expression. The medieval conception of sexual fluid as poison, an artifact of Christianity, is thrown into stark relief when we consider its higher magical usages: the seed of man and woman serves as the impressed *tabula rasa* or magical altar, a bridge between astral and material. Expressions of the corporeal powers referred to as the *kalas* and *ojas* have been written of at length by Kenneth Grant.[11]

Biochemical principles underlying the mechanism of pheromones are of relevance to our path, with particular import to the philter or love-charm, and the herbal disciplines of aromatherapy. Magical rites using such adjunctive materials as incense, anointing oils, and strewing, engage the subtle routes of the olfactory sense and play important roles in ritual atmospherics, and the transdermal absorption of lipidic compounds. As expressed by the body, such may reside in a complex of bodily secretions including perspiration and sebum, the oil expressed

11 *Aleister Crowley and the Hidden God; Cults of the Shadow.*

by the skin. This biochemical agency evokes the philosophical relationships between love and poison; allure, and glamour.

Hormones and steroids, whether calibrated internally through magical technique, or the result of medical therapeutic, may function as a euphoric or dysphoric poison, the former linked with arousal and desire brought to consummation, and the latter from unrequited arousal. At overdose, physicality begins to eclipse both morality and propriety, though an otherworldly dissociative state may also result. This is borne out by a reflective 1998 entry from *Hypnotikon,* after an overdose of intramuscular prednisone, given by a medical doctor for a back injury direct at the inflammation site:

> *The intersection between love and war, so holy among the ancient gods, is borne out in the glands, where the flesh is animated by a 'wordless language.' Here, coitus and violence co-mingle in every fashion, mocking and overturning mortal law. Flesh must conjoin unto itself, for its own glory, and by the extreme routes of the senses, or be torn upon the hooks of Erishkigal. In the arena of the steroidal-gnostic, either event is most pleasing: it is the battlefield aspect of fornication, also allied with all forms of hunting.*

The somatic complex of catacholamines is associated physiologically with high levels of stress, as well as negative external stimuli; they can well be considered 'bodily poisons' in the sense of generating a level of venefic harm. In plants, this group of compounds is a precursor for benzo[c]phenanthridine alkaloids, many of which are currently being investigated as anticancer therapies. Catecholamines also play a role in fear and fight or flight response, which, as a component of magical initiatory process, incepts certain patterns of thought, reasoning, and perception. At such points of crisis, certain matters become possible that were previously closed to all but miracle workers.

These delights and devils, residing in the flesh, are of especial import to operations of sexual magic. Oxytocin is linked with arousal, and prolactin with post-orgasmic euphoria, both states with complex compasses of individual experience and unlimited potentials for magical power. There is also some sexological evidence to support the magical assertion that the higher operations of this art are achieved in dyad rather than alone. Autosexual praxes versus with a partner. Recent research has shown that prolactin release following coital orgasm is increased by a factor of 4 over masturbatory orgasm.[12] The endo-cannabinoids known as CB_1 and CB_2, magically corresponding with the plant drug Cannabis sativa are of note as well, Recent research has shown that CB_1 receptors play a complex role in investigative behavior, curiosity, and response to novel situations.[13] The endogenous cannabinoid anandamide, named for *ananda*, Sanskrit for 'bliss' and 'delight', also plays a role in this complex, as well as pain and analgesia.

The pervasive nature of astral venoms, particular those which generated deformity of the body, was also well mapped in some ancient cultures; the attribution of teratogenesis to good or evil gods was common, as were accounts of star and planet auguries betokening the emergence of the monstrous. Indeed the birth of the limbless, the two-headed and one-eyed, was attended with a disturbing *numen*, and may have affirmed the power of deities with monstrous attributes. Modern science attributes many such deformities to environmental exposure to environmental poison. Where deformities are hereditary, such may have been explained in ancient times as a curse upon a particular bloodline, its first emanation from an evil god or witch.

12 S. Brody S, and T. H. Krüger, "The post-orgasmic prolactin increase following intercourse is greater than following masturbation and suggests greater satiety."

13 Häring, et al. (2011). "Circuit Specific Functions of Cannabinoid CB_1 Receptor in the Balance of Investigatory Drive and Exploration."

Yet, poison's aspect of the Other, or the estate beyond the self, also is implied by the monstrous, as noted by Martin Ruland in his *Lexicon of Alchemy*:

> *Monsters are certain animals which have been originated outside the order of nature, and which heaven hatches in the egg of their putrefaction. The fruit of an unlawful and accursed copulation. They generate nothing in their turn, because they were born from the copulation of diverse seeds, as of a man and brute, or from empoisoned parents. Thus the giants of old originated, and other monstrous men, contrary to nature, whose sight provokes horror and cursing.*

Like the concept of poison itself, mutation may be debilitating or empowering. In essence, it is an indication of the living organism struggling to adapt and persist, even if under duress of vestigial mutation, rather than perish. Implicit, then, is an endemic order of self-perpetuation and diversification. Mutation, foundational to both the theory of evolution and the science of genetics, may thus be regarded as a sacred precept of life, a strategy and armament by which the procession of living bodies radiate over aeons. Poison, as an agent for subtle or gross change, must then adhere to the Paracelsian dicta of relative toxicity. Paracelsus, both a physician and an alchemist, defined the *ens veneni* or influence of poison as one of five such natural emanations acting upon man, with particular reference to its effects upon the gross body.[14] Here it exists within nutriment as a both as an active force and as a potential venom, a kind of malefic *sperma* with toxic potentials.

Such horrors notwithstanding, I have often found that the counterpoison to the sorcerer's 'dwelling in fierce places' is a life of sustained emotional balance, driven by a passion, or love, of the Art Magical itself as one's spiritual consort. Aleister Crowley,

14 *Volumen Medicinae Paramirum*, c. 1520.

himself no stranger to the use of magical poisons, was precise about this:

> *The danger of ceremonial magick—the subtlest and deepest danger—is this: that the magician will naturally tend to invoke that partial being which most strongly appeals to him, so that his natural excess in that direction will be still further exaggerated. Let him, before beginning his Work, endeavour to map out his own being, and arrange his invocations in such a way as to redress the balance.*[15]

The Master Therion's admonition of a personal cartography as a necessary prerequisite to the Art Magical locates the dictum firmly within the dominion of the Magical Body, emphasizing a core Thelemic principle of self-responsibility. The practice of taking stock of one's ætheric body also resonates with the witchcraft Formula of Opposition, one function of which is self-balance and astral homeostasis as the proper vessel for the ingress of power. This principle underlies is exemplified in a number of ritual formulæ, both published and secret, one of which is called 'The Rite of the Opposer.'[16] Such work is essential for the study of magical poisons, and indeed, any magical tradition which admits a complex cosmos inclusive of destructive and mutating powers.

In the years of my work with these pathways I have often encountered the pretender to the Art Magical, who, in misunderstanding this principle, adopts the most obvious and cliché exterior to announce his 'sinister nature' to the world. When such would-be demonolaters approach me and boast of tenebrous fantasies of satanic antinomianism, I do not hesitate to recommend church attendance and sincere acceptance of Christ

15 Magick,'I: The Principles of Ritual.' Note the numerological importance of this section with specific regard to our concern.

16 Chumbley, *Qutub*.

as one's savior. Such statements may seem on their surface extreme, but they are well grounded in history. The deeds of occult orders, even the most sinister among them, simply cannot compete with the morass of two millennia of murder, torture, molestation and repression of the human spirit evident in the *Cultus Christi*. Though Christianity contains a vast abyss of these rank poisons, it also contains, in its higher Gnostic forms, mystical antidotes for its own malediction, such as Christ the Paracelete.[17] It is the Great Work of the Adept of *veneficium* to perfectly reconcile the poison and antidote of the path, whether that thorn-strewn road intersect with religion or not.

It must also be understood that a misperception of the nature of 'Other' frequently arises from the comfortable confines of the Self, be it fantasy or delusion. Though the sorcerer may, by Will and devotion, make himself penetrable to 'otherness', he cannot control the nature of that ravishment, for its power is indeed beyond him. Rather, he can make of himself a Vessel of Transmutation, bearing the powers of flexibility, discretion, cunning, and adaptation, that the illusion of magical bipolarity be broken. Should the basic premise of this be in question, I invite the seeker to keep good company with soldiers, doctors, nurses, and relief workers. Regular encounters with death, disease, and warfare have a means of tempering the soul, and a new relationship with the shadow is born, one which ill tolerates pretension and fantasies. This is dancing upon the point of the magical blade: the conscious embodiment not only of poison, but also of antidote, the venom and nectar of existence.

17 *Christos paráklētos*, the Helper, the Soother, the Healer and Comforter. This aspect of Jesus is known in some forms of Traditional Witchcraft and folk-magic extant in the Cultus Sabbati.

THE DEVIL'S CHRISM

FROM 1994 TO 1998, I PARTICIPATED IN A CLOSED GROUP of practitioners whose sole aim was the magical exploration and development of the so-called 'Witches' Flying Ointment' for the incubation of ritual ecstasis. This salve, sometimes referred to as 'Lifting Balm' or the *Unguentum Sabbati,* occupies a fecund mythic stratum of the occult lore of Old Europe; it has provided a speculative basis, based largely in theories of ceremonial intoxication, attempting to explain the flight of the witches. Fragmentary and speculative formulae from historical sources include the most potent plant poisons in the European pharmacopeia.

The ointment, by custom prepared or provided by the Devil himself, assists the witch in her nocturnal flight to the Sabbat. According to varied reports, the flight is accompanied by otherworldly apparitions, and often includes a series of ritual events corresponding to the features of the Witches Sabbat. Visions encountered by those under influence of the ointment include ravens, owls, bats and other nocturnal creatures; demons, desert wastes, prisons and torments; as well as pleasure-gardens, banquets, jewels, finery, and handsome young men and women.[1]

Interpreting traditional or alleged formulae from early modern European sources, as well as extrapolating historical fragmenta, a considerable number of recipes were developed and tested during our investigation. The approach was multi-tiered and applied vectors of the magical, folkloric, and pharmacological, together with considerations of active folk-magical or traditional ecstatic practices (sustained prayer, chanting, dance, scourging etc.). In addition, we examined lesser-known medical usages of ointments and plasters, to establish parallels and precedents of salve traditions using therapeutic dosages of poisonous plants.[2]

The principal ingredients under investigation using various ratios of formulation were belladonna, opium, aconite,[3] poison hemlock, henbane, poplar, smallage (cultivated and semi-wild celery), and parsley. Additional ingredients based on phylogenetic extrapolation were mandrake, thornapple, angel's trumpet, mugwort, and tobacco. Some of these latter plants were endemic to the Americas, and consideration was also given to historical usage of magical ointments by the Aztecs, as documented by the Spanish.[4] Noting the dermal effects of ambergris, castoreum, ginger, lavender, chamomile, and rose oil, these too were

1 There are also records indicating a belief that the Ointment could destroy crops and interfere with harvests.

2 Such as Unguentum Belladonnæ, historically an official preparation of pharmacy, made under strict laboratory conditions and used for topical pain relief. It is still official in some parts of the world.

3 Two major sources of Monkshood were used: fresh and dried portions of the European *Aconitum napellus*, taken from my own garden, and Oil of Aconite made from *A. columbianum*, a species native to the Rocky Mountains. Some of the work also incorporated prepared *A. carmichaeli* known as Fu-Zhi in Traditional Chinese medicine.

4 Fra. Diego Duran et al, *Book of the Gods and Rites*, 1576. Formulations for these unguents principally consist of venomous animals and the convolvulaceous plant *Turbina corymbosa*.

occasionally included in some measure. Lipidic bases used were beeswax, almond oil, and cocoa butter in varying proportions, although certain animal fats such as lanolin were also used when available.[5]

Adoption of strict protocols of usage early on was critical to our endeavor. Though detrimental effects due to overdose of ointment usage did occur, the more harrowing cases of accidental poisoning took place early on, during preparation of the ointment itself. In the most severe instance, despite extreme precautions taken in handling, I experienced poison hemlock intoxication which was both dissociative and incapacitating. From my magical record of the time:

> *From Hemlock we receive an Education that strangulation, as a phenomenon, is not limited to the region of the neck. The lungs, stomach, muscles, and veins, all experience it and succumb under the fœtid influences of this plant. Indeed, movement, thought, and even self-awareness may also be strangulated... the common element in all cases is slow physical constriction and diminishment of function. Having been so taught, it is unsurprising that our adepts consistently dream of this plant's Genius as ophidian.*[6]

As a student of Traditional Chinese Medicine, Ayurveda, and allopathic herbalism, as well as horticultural botany, my training had well emphasized responsible use of so-called 'folk remedies' as well as given me training in antidote. Beyond my own knowledge and abilities, I was also privileged to work with a number of medical professionals whose advice and talents assisted the operations. Despite the dangers of researching plants of a

5 The increasing availability of African Shea Butter (*Vitellaria paradoxa*) in the United States during the 1990s led me to adopt it as a base parallel to these operations, due to its superior properties.

6 *Hypnotikon.*

moribund and inhospitable nature, the work undertaken in this arena served as a great mentor to my magical disciplines, and a worthy grindstone to my herbalist's blade.

Oils, salves, and magical greases played a considerable role in the earliest documented rites of sorcery as well as religion. The ancient Hebrew predecessor to the apothecary was the *rokeach*, or perfumer, literally "the one who mixes oil with herbs," such were known unto the Romans as *unguentarii*. The association between *maleficia* and magical ointments is also ancient, appearing in the first writings of Sumerian cuneiform texts. *Maqlu* I, 105-16 includes the exhortation '[the witches] washed me with dirty water, they anointed me with a salve made of evil herbs.'

Magic balms and ointments have enjoyed use for a wide range of magical purposes. The *Leyden Papyrus* contains formulae for a considerable number of magical salves, many for the purpose of seeing a god or goddess during invocation; and others used for divination. Still others are intended to provoke the lust of man or woman. A magical ointment for seeing the Bark of Phre appears in the *Leyden Papyrus*, containing such ingredients as ground lapis lazuli, myrrh, 'Great of Amen plant', blood of hoopoe, blood of nightjar, goat's tear, 'Footprint of Isis plant', and ebony wood.[7] Oleaginous preparations were also frequently prescribed for anointing mirrors and other reflective surfaces for scrying operations in which discarnate images were captured in their surfaces. One such mirror, the Mirror of Floron, is found in the 15th century Bavarian *Munich Handbook*. It is constructed of gleaming steel and smeared with "pure and bright balsam" prior to use.[8] A similar usage seems to have been associated with Sa-

7 Griffith and Thompson, *The Leyden Papyrus*. p. 81.

8 Richard Kiekhefer, *Forbidden Rites*, pp. 104–105. Other references abound through the work, such as olive oil used to anoint a mirror, p. 106. Balsam is also used in old magical and herbal manuscripts to indicate oleoresins such as frankincense, the plant lemon balm (*Melissa officinalis*) or healing substances in general.

dyngstone, abbot of Leicester, in 1440.[9] In the British Isles, there are records of magical ointments for the eyes compounded of four-leaved clovers. This balm enabled the sorcerer to see into the land of the fairies, whilst watching the lands of humans disappear, a significant occurrence of magical salve because of its reality transmutation despite lack of 'classical' Sabbath references. The famous medieval manuscript *Secrets of Albertus Magnus*, includes the following formula:

> *If thou wilt see that other men cannot, take of the gall of a male Cat, and the fat of a Hen, and mix them together, and anoint thy eyes, and thou shalt see that others cannot see.*[10]

Mallow (*Malva* spp.) ointments were traditionally used to remove hexes from bewitched persons, as well as to protect a person from the ravages of hot metal;[11] uses which, given Mallow's demulcent medicinal properties, have a sound pharmacological basis. In certain regions of India, ointments containing lampblack are smeared on the eyelids to protect against the Evil Eye, or to frighten away noxious spirits. The diffusion of European ointment-based sorcery into the Americas often found a willing consort in endemic magical practices and a poison-rich flora. In the folk magic of Virginia, an ointment made from foxglove and rattlesnake oil was prepared to rid the effects of being blasted.[12]

Considering the ointment from a pharmacological perspective, the existence of an hallucinogenic or ecstasy-generating salve is both possible and probable. As practicing medical herbalists know, highly concentrated plant virtues in lipidic bases have a venerable history. In Roman medicine, Pliny the Elder reports

9 Op. cit., p. 100. Through such actions the abbot allegedly sought, and discovered, a thief guilty of robbing the treasury.

10 From the chapter 'Marvels of the World.'

11 Charles M. Skinner, *Myths and Legends of Flowers, Trees Fruits and Plants*, p. 168.

12 Hubert J. Davis, *The Silver Bullet and Other American Witch Stories*, p. 161.

that Oil of Henbane was made by steeping the plant in olive oil and pressing. *M'shihat popilion* is an ointment mentioned in Jewish alchemical texts, to which Raphael Patai appends this footnote:

> *In modern Spanish,* populeon *is a popular ointment prepared with pigs' butter, poppy leaves, belladonna, and other things, and used as a sedative. Its principal bases consist of the first shoots of black poplar leaves.*[13]

Pharmacopeias of the 1700s and 1800s provide clear instructions for making anodyne ointments from *Atropa belladonna.* At the time of my early research into these matters, use of a concentrated Oil of Aconite for a foot injury led me to investigate greater levels of application beyond therapeutic analgesia; with edifying results. Where established 'safe' levels of concentrated medicines are calibrated for most human physia, it is often a short step from 'therapeutic' dose to 'overdose', though the latter can be quite variable, having minor or severe symptoms.

The late medieval physician Johannes Hartlieb is credited as providing the first written description of the witches' flying ointment.[14] He speaks of the *unguentum pharelis,* an ointment compounded of seven plants, blood of a bird and animal fat. This was used to anoint chairs and other objects upon which sorcerers would ride.[15]

Six years before Hartlieb's description, another ointment recipe associated with the Witches' Sabbat appeared in the tractate *Errores Gazariorum seu Illorum qui Scobam vel Baculum Equitare Probantur,* penned by an anonymous inquisitor. Though it is not

13 Raphael Patai, *The Jewish Alchemists.* p. 566.

14 This attribution is largely confined to the historical theatre of the European witchcraft heresy. A better ancient prototype resides in Apuelius's *The Golden Ass,* where Pamphile is transformed into a screech owl by aid of an ointment and muttering charms.

15 *The Book of All Forbidden Arts,* c. 1455.

associated with night flight, the salve contained toads, snakes, lizards, and spiders, mixed with "a diabolical mystery, a touch of which kills, either with a sudden or lingering death". This ointment is situated amidst one of the most noxious and scandalous descriptions of the Sabbat, including repeated diabolic uses of the intestines of children.[16] Describing the ointment in a form which is closer to the modern understanding of it, the Franciscan jurist Alonso de Castro wrote:

> *There are certain oyles and oyntments with which they anoynt themselves, which deprive them of their right sense, making them imagine they are transformed into birds or beasts, deceiving not only themselves with this error, but oftentimes the eyes of other, for the devil and other enchanters so deceive our sight, turning and transforming men into beasts to the seeming of those which behold them, though in truth it was nothing so, but the sorcerers think themselves in their imagination to be transposed. Sometimes they anoynt themselves with other oyntments whose operation maketh them think they are like fowls and can fly into the air.*[17]

This reference takes a skeptical approach to the witch's power, portraying the ointment as an agent of delusion, a stance functionally resembling that of the Natural Magician. The passage is important as it documents the Ointment as a known phenomenon, even as it is repudiated as a mundane influence. The simultaneous admission of the salve's existence, but dismissal of it as a delusory trick, is reminiscent of the testimony of Chonradt Stoecklin, a Bavarian horse-wrangler tried for witchcraft in the late sixteenth century. During his trial, he denied use of the oint-

16 Lea, Henry Charles. *Materials Toward a History of Witchcraft*, I:274.

17 Thompson, C. J. S *Mysteries and Secrets of Magic*, p. 121.

ment for nocturnal flight and "claimed that he needed no artifice for his journey."[18]

In his *De la démonomanie des sorciers* of 1580, the French jurist Jean Bodin presented the following anecdote:

> *Paul Grillandi, who lived in 1537 relates that a certain Sabine, living near Rome, was persuaded by his wife to rub himself with a certain ointment while she recited certain words, so that he might attend a witches' Sabbat. He suddenly found himself under a huge nut tree in Beneventum with a large assemblage of witches carousing. He did as they did until he called for some salt, which is very repugnant to the devil... at the mention of the sacred name, the whole assemblage vanished, and he found himself lying naked 100 miles from home. He had to beg his way back; and, recognizing the impious character of the proceedings, which his wife had concealed from him, he accused her and she confessed and was burnt, along with a number of her accomplices, whom she pointed out.*

Actual use of a mind-altering ointment in the context of European witchcraft is controversial among historians. Certain scholars of the history of European witch-persecution have taken a skeptical or dismissive position with regard to the existence or use of the ointment,[19] whilst other writers, mainly in the sphere of 'entheogenic research' treat it as a historical certainty.[20] Both positions, as well as others, tend to ignore evidence that suggests the phenomenon was wider than both magic and pharmacology, and more widely distributed through time than the period of witch-persecutions.

18 Behringer, *Shaman of Oberstdorf*, pp. 93–4.

19 See, for example, Briggs, *Witches and Neighbours*, p. 56; Henningsen, *The Witches' Advocate*, p. 391.

20 Christian Rätsch, *Hexenmedizin*.

What is certain is that, as has been shown, a large body of *lore* exists concerning this ointment, in both the written historical record and in the popular imagination. The fascination it has exerted over practitioners of folk magic, themselves the keepers of occult records and oral teachings, has persisted for centuries. Rural magical practice participates in its own body of lore and custom with an investment of legitimacy equal to or greater than what exists in the written record, a verbally transmitted 'cult of ancestors' which, despite their status as 'ideas', act and behave in the context of magical practice as informing spirits. As such, custom and lore represent a complex magical cipher which, when rooted in the flesh of the initiate, offers private revelatory dimensions specific to the individual path.

The relevant question may not concern actual historical existence of a psychoactive ointment, but whether or not it was used in the context of that popular magic coming under the lens of the authorities as 'witchcraft'. Though most modern research into the Sabbatic Ointment plumbs legal and inquisitorial literature of the witchcraft persecutions, its presence outside this context, such as abides in collections of folklore or contemporary histories, indicates that it was clearly an important *topos* present in European magic. Folklore of Belorussia and the Ukraine relays that naked witches flew up chimneys, naked and covered with ointment, on broomsticks or pokers, and onward to a *Shabash* on Bald Mountain near Kiev. The flight was aided by an asperge of water boiled with ashes from the midsummer fires, or an ointment made from Gentian.[21] In at least one case, an accused witch mentioned a magical ointment provided by the Devil at the Sabbat, not for flying, nor as an inversion of Christian ritual, but to heal the wounds on the back of a celebrant recently whipped with branches of thorn-bearing *Crataegus*.[22]

21 W. F. Ryan, *The Bathhouse at Midnight: Magic in Russia*, p. 80.

22 Gustav Henningsen, *The Witches' Advocate*, p. 155.

In the course of our experiments, formulations of the salve differed, but method of application and ritual procedures were similar, even as they evolved over time. There was also present, in each case, the so-called 'watchful companion', a participant not anointed or in an otherwise compromised physical state. This feature of our work was at my insistence, knowing the treacherous nature of solanaceous alkaloids, but also for the benefit of dual perspective on anointing events. The resulting data collated over many operations afforded a rich diversity of perspectives from practitioners with widely varying sensitivities and constitutions.

In addition to more 'traditional' formulations, salves were also compounded of plants presenting lesser toxic hazard. When tested, many provided edifying results. For example, the role of aromatic features, not suspected before the work began, was essential in forming 'Ground State Calibration.' When a certain olfactory key, such as Attar of Rose or Oil of Rosemary, was provided before the anointing proper, it could be used for a 'sensorial return' or 'grounding' at the appropriate time, much like a hypnotic cue, and became a valuable part of the procedures.

In another instance, a lipidic sexual ointment consisting of extracts of mugwort and wormwood was found efficacious in assisting the trance state as well as vastly expanded pansomatic tactile sensitivity. However, due to its tendencies to stimulate uterine contraction, its use was not engaged by pregnant women. A relevant entry from the *Hypnotikon*:

> Anointing Rite: *15 April, 1995—as transcribed from operants DH in the station of Master, and DG in the station of Mistress. Procedure: Sustained ablutions, exorcism of three creatures,*[23] *LBRP,*[24] *Lunar propitiations as usual to Diana and the noc-*

23 The exorcism-triplicity of Fire, Water and Salt, as per Solomonic grimoire formulæ.

24 Lesser Banishing Ritual of the Pentagram, Crowley, *Magick*.

turnal company en masse, in their prayerful and sigil-based forms. As discussed last meeting the Blessing of the Oil was performed in the more fully developed form—and called as the presiding angel of the rite.

Pattern of Anointing:

1. Soles of feet & between toes
2. Behind knees
3. Perineum & base of spine
4. Phallus/Kteis—external only
5. Belly
6. Elbow & juncture of forearm & upper arm
7. Neck
8. Behind ears & Temples
9. Forehead
10. Crown

Oil formulation: *'Deviant-1' so named because it is the first to include non-traditional constituents, in this case* oleum essentia Artemisia[25] *and oil of hashish as distilled and kindly provided for our purpose by G. Application at each point on the body was a sustained massage with slight pressure. Feelings of exhilaration ensued within the first hour accompanied by slightly blurred vision and increased tactile sensitivity. A second anointing followed via Agapæ,[26] during which sensations of floating were reported with an onset of intrusive coloured shapes and 'sparkles' in the vision. Both floating sensation and visual effect persisted (and increased) through orgasm and*

25 Essential oil of Mugwort (*Artemisia vulgaris*).

26 The sexual offering.

> *into refractory phase, and into subsequent dream during sleep. Despite the fact that both participants focused on the tactile and visual sensations, the report of floating and 'being outside oneself' remains crucial to our inquiry. Sexual activity proceeded despite risk of overdose or irritation via transvaginal/transpenile absorption; on the contrary, these two participants reported only pleasurable sensations.*

Modern occultists and drug enthusiasts persist in the assumption that the medieval and early modern European witch applied the Unguent to the broomstick and anointed the genitals by either prolonged ritual masturbation or actual insertion, and thus 'rode the broom.' Beside the fact that such assumptions are historically speculative and oversimplify the situation with regard to actual magical practice, this ignores the witches' broom as part of a much larger historical corpus of flying mounts of the night- wandering Host, many of which are non-phallic.[27]

Nonetheless, in the operations of which I speak, specific routes of administration included the 'secret place', precisely because of the perpetuation of the genital-anointing myth. In the experiments, ritual anointing included massage or rubefaction, with especial attention to the feet, genitals, palms, neck, and scalp. The more intrepid participants in the cabal used the highly toxic and concentrated salves as sexual lubricants, an action which violated ritual protocol and put their physical well-being at risk. Results of these formulations via this route ranged from mild stimulation to a partially dissociative somatic ecstasy which mirrored in some respects the alleged features of the witches' nocturnal flight. Though this route of administra-

27 Carlo Ginzburg, *Ecstasies* pp. 163–164. Included in this retinue of airborne objects are scythes, benches, and bowls. The Abkhaz sorcerers flew through the sky on uprooted trees, cart-wheels, oven shovels, and other objects. Similarly the Circassian sorcerers were borne aloft upon fishing boats and animal carcasses, pieces of which occasionally rained down as they flew.

tion makes certain alkaloids more physiologically bioavailable, it is also highly dangerous. This evokes the related practice of 'keistering', whereby a psychoactive drug is inserted anally or in some cases vaginally, which has ancient antecedents.[28]

It is also of interest to note that usage of our prepared ointments in sexual or magico-sexual contexts gave rise to unique instances of enhanced tactility. Whilst these experiences in no way resembled flight, it would be fair to report an expansion and elevation of the sensorium falling under the general categories of a stimulant and exhilarant. Given that the salves employed mugwort, their energetic dispensation was reminiscent of reports of an Aztec ointment made from a Mexican wormwood species, *Artemisia mexicana,* and listed as a 'Remedy for Lassitude' and rubbed into the soles of the feet. More intimately and to the point, the Kama Sutra contains a number of magical ointments, some of which feature the same poisonous plants of the nightshade family associated with the European witch-salve. One such preparation, employed to bewitch a woman, anoints the *linga* with a mixture of ground thorn-apple seeds and honey. When the man goes to her, she will be 'caught in his power.'

One obstacle in approaching the matter of the *Unguentum Sabbaticum* arises in the common assumption that psychoactivity is the function of a singular cause and effect. To those who research the Ointment, especially from the historical perspective of examining extant written records, it quickly becomes apparent that we are accessing the corpse of a magical tradition, and an incomplete one at that. Names of plants differ, or are sometimes imprecise, proportions vary (if given at all), and a number of ingredients appear which the pharmacologically-inclined tend to dismiss as 'sympathetic magic.'

In a presentation given on the subject in 2007, I asserted that one of the greatest barriers to understanding the witches' ointment is the occlusion generated by a purely ethnopharma-

28 Peter DeSmet, *Ritual Enemas and Snuffs in the Americas.*

cological perspective.[29] The same is true in consideration of anthropological or micro-historical inquiry. These limited approaches become even more unreliable considering the challenges presented by the secret nature of magical practice and the toxicity of the plants involved. Until the needful perspective of the magical practitioner is considered and applied as a lens to the matter, and from multiple angles, holistic comprehension of the subject will be elusive. This added perspective necessarily includes an adjunctive field of supportive magical praxes, nuances of intent, lore and conceptions of the supernatural, cultic symbolism, magical fetish objects, and alternate, non-pharmacological routes of attaining deep ritual trance. The importance of this latter consideration is underscored in a statement by Éva Pócs, whose career has focused on supernatural belief in early modern Europe, including witchcraft:

> *The subject of flying ointment occasionally came up in Hungarian interrogations; nonetheless, it played no real role in precipitating trance. In our trials, no evidence underpins the assumption of some European research that witches, including Hungarian witches, made journeys with the help of drugs. Documents referring to altered states of consciousness clearly talk about spontaneous trance. References to flying ointment are generally traceable only in responses given to questions of the court, and even then principally in confessions following torture. Flying ointment appears in the narratives as a metaphor for creating trance, in the context of flying with wings—that is, "really" flying—or in the legendary motif of turning into an animal.*[30]

'Non-pharmacological' routes of spirit flight, many involving ritual induction of trance, are also documented contemporary with

29 Ludlow Esoteric Conference, Shropshire, June 2007.

30 Pócs, Éva. *Between the Living and the Dead*, p. 77.

reports of ointment usage. A particularly interesting example is reported from the *Charovnik*, an old Russian book of spells no longer extant. The text gave instructions on 'how to leave your body as if as if dead and fly like an eagle or a hawk or a crow or a magpie or an owl, or run like a panther , a savage beast, a wild beast , a wolf or a bear, or fly like a serpent.' W. F. Ryan notes that while most of the banned Russian magic-books were of Byzantine origin, but the *Charovnik* has clear Russian shamanistic features.[31]

Co-factors present in consideration of ritual ointment usage include the notoriously variable arenas of 'set' (the estate of the individual) and 'setting' (the estate of the immediate environment) first described and categorized by Timothy Leary. However, because of the status of the psychoactive unguent as imbedded within a magical context, 'set' of necessity encompasses magical intent, protocol and taboo, and attachment to outcome. Elements of 'setting' include such adjunctive ritual techniques as bathing the body to open pores of the skin, the better to facilitate absorption; massage or rubefaction, for increasing circulation; movement or sexual activity, breathing exercises, withdrawal and isolation. Finally, perhaps the greatest consideration of setting is the virtues and character of the plants themselves, either singly, or as a magically-bound troop fixed within an oily base.

This leads us to consider the nature of the so-called 'Hexing Herbs', and particularly solanaceous intoxication. One characteristic of tropane inebriation is a distortion of time, not only in its perception by the user, but also shifts in temporality itself as an embodied feature of the narcotic Sabbat. The procession of Time, its acceleration, deceleration, fragmentation, and defining effect upon the visionary arena, is often as important as the interplay with the spirits themselves. This bears obvious corollaries with 'Fairy Time' in Brythonic lore, wherein humans abducted into the realm of Faerie attend a great feast in the sub-

31 W. F. Ryan, *The Bathhouse at Midnight*, p. 55.

terranean halls of Elphame, only to discover upon their return to the world of men that many years have passed, as with the legend of King Herla. In addition to the strong dissociative tendencies, this quality of nightshade poisoning marks it as especially dangerous; for this reason I have referred to the visionary tropanes in previous writings as *chronophagoi* or 'time-eaters.'[32]

Additionally, the perpetuity of effect, or an oscillating tendency of emergence from, and sudden return to the delirium, reveals in these plants the presence of an insidious sentience transcending the parameters of chemistry. This is dramatically illustrated by a tale recounted to me by a fellow practitioner some years ago, who at the age of 12 consumed an excess of *Datura stramonium* seeds. After a harrowing three days of navigating a wasteland filled with walking corpses, he seemed at last to gain waking respite near a sunny meadow, becoming aware of the sober individuals who had been his caretakers during his break with reality. His gaze shifted to a grassy swale, wherein he glimpsed a flittering. Amid the reeds a watchful satyr stood, observing him intently. Maintaining his gaze, the silen slowly raised what appeared to be a police radio to his lips and whispered into it:

> *"Affirmative. We have one who sees."*[33]

During the years in which the ointment research took place, I collected several first-hand reports of western travellers in Indonesia who were drugged by thieves. This occurred to such an extent, with such consistency of symptoms, that it in a short time it became a known phenomenon to be vigilant for among wayfarers. In all cases I collected, of greater personal horror to the victim than the crime itself was the state of waking catatonia brought on by the drug, combined with fear, motor paralysis, and

32 Schulke, *Viridarium Umbris*, 2005.

33 Many thanks to Robert Fitzgerald for permission of reference.

monstrous hallucination, followed by blackout. These adumbrations of the grotesque easily match tropane intoxication, and indeed in one case the victim was told by a local that it was a known thieves' grift, of which *Datura* was a favored ingredient. Tropane narcosis easily conforms to the 1507 description of the witches ointment by Alfonso Tostado, in his *Super Genesis Commentaria*:

> *there are women who we call* maleficæ *in Spain who say that with an ointment and certain conjurations they are carried to distant places where there are assemblies and enjoy all kinds of pleasure. But this is an error, and has been found by their falling into a stupor, insensible to blows and fire, and on awaking in a few hours relating where they have been and what they have seen and done.*

Tropane alkaloids have strong dissociative effects, as well as a platform of other strong narcotic phenomena, which can also lead to death. Whilst dosage is clearly a factor, individual body chemistry and psychology are strongly suspected as contributing to the degree of intoxication. Tales such as I have relayed represent the testimonies of the lucky, who were not only able to survive physically, but also recover memories of it. Such accounts, their grotesquerie notwithstanding, present an amalgam of the daimonic and the projections of the mind. One the one hand, the settings and language expressed are modern, but on the other, the presence of satyri and kindred demonic forms participates in the well-known solanaceous generation of phantasmagoria, an essential component of the iconology of European witchcraft. Yet such tales must also be responsibly viewed against the greater backdrop of misuse of the visionary tropanes, wherein loss of sanity, limbs, or life has occurred. From my final 1998 entry on this subject in the *Hypnotikon*:

> *I am present at the Assembly. All gathered there are beautiful—each possesses a noble countenance. There is a warmth*

present which seems not only related to temperature but also to our mutual felicity. The room was prepared simply, but increasingly acquires a rich and densely-layered ambience, as a grand feasting-table laden with every kind of food and drink. Macropsia is random and extreme: candles burn in their assigned stations; at times they suddenly flare much larger, like small bonfires, then just as suddenly return to a small flickering. Each visual part of the present experience is divided an infinite number of times, recursive and returning upon itself. I am disturbed to discover that each thing demands to be experienced singly, to the exclusion of all others: a face, a flame, an object lying in shadow. Singular focus gives each object sense of the extruded and apparitional. Attempts to experience the integrated whole as a unity increasingly cause violent discomfort and a disembodied floating sensation. At such times, the beauty of each 'scattered portion' becomes visually horrid. The idyll is punctured by the apparition of large curved hooks, like great scythes or claws, descending downward, tearing the scene before me like stretched and painted hide. Behind it lies skeletal rottenness and desolation, the death of color, bleak wastes extending to a distant horizon, and an endless sensation of falling. Loud winds rush violently into the chasm, attempting to extract all remnants of the formerly pleasant tableau into its distant extremities. A certainty dawns on me that a terminus has been reached: there will be no return, only an ever-increasing velocity Beyond.

GLOSSARY

æromancy: divination by clouds.

alchemism: magical process which, although not classical alchemy, nonetheless retains its central feature of transmutation of one essence to another.

alraun: magical fetish made from the root of *Mandragora.*

Arena of Power: a concept arising from Sabbatic Witchcraft, but applicable to all schools of magic and mysticism, describing the universe of the sorcerer as an infinitely dynamic source of initiatic consciousness and *numen*.

Astral Poison: a poison that afflicts, corrupts or destroys the spirit.

atavism: shade of pre-incarnate existences, summoned using magical formulæ or appearing unbidden as an intrusion.

azotic: of or pertaining to Nitrogen, its chemical sentience, and its compounds. Nitrogenous compounds are often concurrent with psychoactivity, as with the alkaloids, though there are a few rare non-nitrogenous exemplars.

corporeal laboratory: a magical model of the human body used for magical transmutation.

Crooked Path: division of magic describing a unified ethos of blessing and cursing, or right and left-hand paths of magic, present in a number of systems of sorcery throughout history, but most notably in the magic and lore of the witch-mythos. Masked in the glyph of the Serpent, its wending nature betwixt all points of universal power precipitates initiatic consciousness. The classic texts of Crooked Path Sorcery are *Qutub, or the Point,* and *The Dragon Book of Essex* both by Andrew D. Chumbley, and *Via Tortuosa* by Robert Fitzgerald and Daniel A. Schulke.

crow's bread: ritual plant poisons used to secure mystical or visionary experience. The name originates in Essex Traditional Witchcraft and can also apply to the mushroom *Psilocybe semilanceata*, and, less frequently, other plants such as *Atropa belladonna*.

distillation: the work of separating and potentiating the virtues of a substance by way of controlled heat and condensation.

Edenic Body: the 'Corporeal Garden' of endogenous poisons and psychoactives, and their magical potentials.

emetic: an agent which causes vomiting.

gnostic poisoning: attainment of spiritual knowledge via intoxication.

graal: grail, magical cup, or its contents.

hexenrauch (witches' smoke): in German witch-lore, a witchcraft suffumigant or incense.

hypercapnia: condition of CO_2 intoxication in which there is too great a quantity of carbon dioxide in the blood.

LD-50: in older schools of toxicology, 'lethal dose 50%' or median lethal dose; the amount of a substance to kill 50% of the test population (now obsolete).

maleficia: magic operations for the purpose of harm. Contrasted to *beneficia*.

mithridate: a universal specific against poisons, named for Mithridates VI Eupator, ancient King of Pontus.

mumia: in Paracelsian alchemy, the liquor derived from distillation of the human body, in particular one which has died a sudden death, and thus 'harvested' whilst still alive and healthy. In Sabbatic Witchcraft, mumia refers to human bones, blood, flesh, as well as sexual fluids, as used in the Magical Art, and also the power or essence residing therein, cognate in some sense with the Yoruban *ashe*, or Pacific Islander *mana*.

narco-aesthesis: induction of a mystico-ecstatic state via the ritual use of drugs.

Narcotic Sabbat: the medieval witches Sabbat as speculatively arising from the usage of psychoactive plant, animal, and mineral substances.

New Flesh: mystical concept present in Sabbatic Witchcraft relating to magical augmentation of the Self, drawing upon ecstatic ritual, atavistic power, and the sorcerous formulation of Will, Desire, and Belief. Its magical attainment shares certain characteristics of the general magical purview of 'shape changing.'

Opposer: spiritual agency of inversion and opposition, which in some forms of English folk magic is cognate with *devil.*

'other': powers, knowledge and experience alien to Self; the shadow consciousness or Abyssal Self.

philtron: in the love-magic of ancient Greece, a love-potion. Depending on context, *philtra* may also denote love-spells.

phytognosis: magical knowledge gained via the agency of plant powers.

rubefaction: rubbing or scourging of the skin to increase blood circulation. In certain schools of Alchemy, allied to *Rubedo,* or the reddening, the name given to the final phase of the work.

Sabbat: From the Sumerian *sabat-tu,* 'Day of the Heart's Rest', it is the primal ritual patterning of the Gnosis of the Witch is the Rite of the Sabbath, known also by the names of Covenant, Convocation, and the Great Congress. Often regarded as a degraded relic of archaic Babylonian high magic, or as a fantasist construct of the Christian pogroms against heresy, it has passed through both these aeonic phases intact, and subsumed what is necessary for its continuance. Its magical recension in the present era is Sabbatic Witchcraft.

Sabbatic Witchcraft: a modern corpus of orally-transmitted ritual magic techniques, spells, lore, and concepts ultimately deriving from the medieval witches' sabbath. The name was coined by Andrew Chumbley to describe the tradition of Buckinghamshire-Essex Craft into which he had been inducted, previously nameless, but which now refers to itself as *Cultus Sabbati*. Although containing teachings and practices of previous historical eras, it continues to evolve and incorporate material of the present, based on each successive generation of initates.

saturika: in the ancient Greek pharmacopeia, a designation used to indicate erection-producing drugs.

Sexual Poison: a sexual emanant, whether astral or fluidic, which has been subjected to the Formulæ of Opposition such that its life-giving virtue has become apostate unto itself.

solanaceæ: plant family of toxicological interest including tropane-bearing species such as Belladonna and Mandrake.

strewing: the ritual act of casting fragrant botanical materials in the ritual area, often as a sacramental offering or serving the function of consecration. The area thus strewn may be a ritual boundary, as in the magician's magic circle, a votive area or shrine, or a pathway to be trodden, often barefoot. Traditional strewing plants thus used are lavender, chamomile, and rose, but any fragrant plant, or combination thereof, may be employed.

suffumigation: magical incense, especially in the literary context of the European grimoire; also *perfumes*.

theriac: an agent which prevents poisoning, often taken as a tonic or cumulative.

thrombolytic: an agent for dissolving blood clots.

titration: in medicine, the modification of dosage by gradual increase or decrease for optimum pharmacological effect.

tonic: an agent which gradually strengthens over time with regular use.

transmigration of poison: concept of the movement of poison from one body to another, and its accompanying evolution.

Transmutation of Poison: Operations of magic whereby poison becomes nectar within the Vessel of the Practitioner. In Crooked Path Sorcery, this operation is a form of 'Self-Overcoming' or destruction and reformation of self as a dual manifestation of the Path and the Secret Initiator.

Tubal-Cain: According to some, the first alchemist; according to the Bible, the 'instructor of every artifice of brass and iron' and a descendant of Cain. Historically, he occupies rolê of prominence in a number of concealed lodges of Traditional Witchcraft.

Verdelet: 'Green Master' of a witches' covine, responsible for teaching and stewardship of herb-lore.

Vinum Sabbati: the mythic wine of the Witches' Sabbath, according some legends possessing magical or intoxicating properties. A *Lapis Lamiis* or [Philosopher's] Stone of the Witch, it is emblematized as 'The Wine of the Sabbath', but may take the forms of a vintage, a transmission of power, and a spirit-emanant of the flesh.

Witches Supper: feature of the medieval Witches' Sabbath portrayed as ritual cannibalism, but interpreted by some as the ritual use of psychoactive sacraments.

Witching Cup: a magical component of some forms of Traditional Witchcraft wherein a ritual draught is administered. The formula of the draught may be psychoactive, either of plant, mineral, or animal origin, including human. In some traditions this may take the form of a 'Mock Communion' or blasphemed Christian sacrament; such may also include the 'Wine of the Sabbath' (*Vinum Sabbati*) of the Sabbatic Current.

wort: a plant, particularly a medicinal herb that is not a tree. Alternatively, wort may refer to the cooking brew of malt, water, hops, and other herbs that, after being cooled and fermented, becomes beer.

Acknowledgements

The author expresses his sincere appreciation to the following individuals and groups who have stood as companions of the poison convivium. Dale Pendell, Constance D. Binero, Jack Darkhand, Peninsula Entheogen Research Group, Greg G., Andrew D. Chumbley, Robert Fitzgerald, the women of the Amber Besom Coven, Rabbit, Deadly Nightmare, Laurie Kaufman, Liz Spedding, John Wheallan, the Crooked Rabbi and the host of *shaitans*, Cody Dickerson, Billy Davis, and Arbor Infernis Cell.

BIBLIOGRAPHY

Adamse, P. and Van Egmond, H. P. "Tropane Alkaloids in Food." *Method Development and Analysis of Mycotoxins and Phycotoxins in Food*, RIKILT, Institute of Food Safety, Wageningen University & Research centre Akkermaalsbos 2, 6708 WB Wageningen, the Netherlands, 2010.

Agrippa, Henry Cornelius of Nettesheim. *Three Books of Occult Philosophy* [Donald Tyson, ed.] St. Paul: Llewellyn, 1993 (Antwerp, 1531).

Ankarloo, Bengt and Stuart Clark, eds. *Witchcraft and Magic in Europe: Biblical and Pagan Societies.* University of Pennsylvania Press, Philadelphia, 2001.

Ballantyne, John William. *Teratogenesis: An Inquiry into the Causes of Monstrosities.* Oliver and Boyd, Edinburgh, 1897.

Baumann, Helmut. *The Greek Plant World in Myth, Art, and Literature.* (*Die griechische Pflanzenwelt in Mythos, Kunst und Literatur*) trans. William T. Stearn and Eldwyth Ruth Stearn. Timber Press, Portland Oregon, 1993 (Munich, 1982).

Behringer, Wolfgang. *Shaman of Oberstdorf: Chonrad Stoeckhlin and the Phantoms of the Night.* University Press of Virginia, London, 1998.

Betz, Hans Dieter, ed. *The Greek Magical Papyri in Translation* (2nd ed.). University of Chicago Press, Chicago, IL, 1992 (1986).

Bord, Janet. *Cures and Curses: Ritual and Cult at Holy Wells.* Heart of Albion Press, 2006.

Boot, Christine: "Home and Farm Remedies in a German Manuscript from a Texas Ranch," in *Paisanos: A Folklore Miscellany*, ed. Francis Abernethy, 1978.

Braude, Pearl F. "Cokkel in Oure Clene Corne: Some Implications of Cain's Sacrifice." *Gesta* vol. 7, 1968.

Briggs, Katharine. *British Folktales*. New York; Pantheon Books, 1970.

Brody S, and Krüger T. H. "The post-orgasmic prolactin increase following intercourse is greater than following masturbation and suggests greater satiety." *Biological Psychology*, 2006 Mar; 71(3): 312–5.

Bruno, Girodano. *On the Composition of Images, Signs and Ideas*. Trans Charles Doria. New York: Willis, Locker and Owens, 1991 (1591).

Budge, E. A. Wallis. *The Divine Origin of the Craft of the Herbalist*. Dover Publications, Mineola, NY 1991 (London, 1928).

Camporesi, Piero. *Juice of Life: The Symbolic and Magic Significance of Blood*. Continuum: New York, 1995.

———. *Bread of Dreams. Polity Press, Cambridge, 1989.*

Christian, Paul. *The History and Practice of Magic* (*Historie de la Magie, du Monde Surnaturel et de la Fatalité à travers les Temps et les Peuples*), trans. James Kirkup and Julian Shaw). New York, NY: Citadel Press, 1969 [1870].

Coffee, Timothy. *The History and Folklore of North American Wildflowers*. Facts on File, New York, NY 1993.

Coffin, A. I. *Botanic Guide to Health*, and the Natural Pathology of Disease. British Medico-Botanic Press, Manchester, 1850.

Conjuration for Magitians and the More Learned Sort of Conjurors, Society for the Esoteric Endeavour, 2006.

Chumbley, Andrew D. *Azoëtia: A Grimoire of the Sabbatic Craft*. Chelmsford: Xoanon, 1992 (2002).

———. *"The Secret Nature of Ritual" Chaos International No. 18, 1995.*

———. *"Gnosis For the Flesh Eternal" Opuscula Magica 2, Three Hands Press, 2011.*

Crowley, Aleister. *Magick-Liber ABA, Book Four, Parts I–IV*, Samuel Weiser, York Beach, Maine, 1994.

———. *The Goetia. [Mather, S.L.MacGregor, trans.] Samuel Weiser Samuel Weiser, York Beach, Maine, 1995.*

DeLancre, Pierre. *On The Inconstancy of Witches* (trans. Gerhild Scholz Williams). Arizona Center for Medieval and Renaissance Studies, 2006 (1612).

DeLanda, Manuel. *Uniformity and Variability*, Zero News Datapool.

deWaal, Dr. Marinus. *Medicines From The Bible*. Samuel Weiser, York Beach, ME, 1984.

Duerr, Hans Peter. *Dreamtime: Concerning the Boundary Between Wilderness and Civilization*. New York, NY: Basil Blackwell, Inc. 1987. Faraone, Christopher A. *Ancient Greek Love Magic.* Harvard University Press, London, 1999.

Evans, George Ewart. *Ask the Fellows who Cut the Hay*. Faber and Faber, London, 1965.

Fanger, Claire, ed. *Conjuring Spirits: Texts and Traditions of Medieval Ritual Magic*. Pennsylvania State University Press, University Park, PA, 1998.

Ficino, Marsilio. *Book of Life* (*Liber de Vita*), 1489. [Trans. Charles Boer; Spring Publications, Woodstock, CT 1994].

Filotas, Bernadette. *Pagan Survivals, Superstitions, and Popular Culture.* Pontifical Institute of Medieval Studies, Toronto, 2005.

Fitzgerald, Robert. *Midnight's Table*, privately published.

Goetz, Delia, and Morley, Sylvanus G., trans. *Popol Vuh: The Sacred Book of the Ancient Quiché Maya.* University of Oklahoma Press, Norman, OK, 1969.

Gupta, Shakti. *Plant Myths and Traditions in India.* Leiden, E. J. Brill, 1971.

Grant, Kenneth. *The Magical Revival.* Samuel Weiser, York Beach, Maine 1972.

———. *Aleister Crowley and the Hidden God. Skoob Books Publishing, London, 1973 (1992).*

———. *"Vinum Sabbati" Carfax Monographs. Re-issued as Hidden Lore, Fulgur Limited (date).*

—*Nightside of Eden*, London, Fredrick Muller, 1977.

Haggard, Howard W. *Devils, Drugs, and Doctors*, Blue Ribbon Books, 1929 p. 301.

Häring, M.; Kaiser, N.; Monory, K.; Lutz, B. (2011). Burgess, Harold A.. ed. "Circuit Specific Functions of Cannabinoid CB_1 Receptor in the Balance of Investigatory Drive and Exploration". PLoS ONE 6 (11): e26617. doi:10.1371/journal.pone.0026617. PMC 3206034. PMID 22069458.

Herodotus. *The Histories.* Trans. Robin Waterfield. Oxford University Press, New York, 1998.

In Praise of Ale: Songs, Ballads, Epigrams and Anecdotes Relating to Beer, Malt and Hops. George Redway, London, 1888.

Jacquart, Danielle and Claude Thomasset. *Sexuality and Medicine in the Middle Ages.* Princeton University Press, 1988.

Jolly, Karen Louise. *Popular Religion in Late Saxon England: Elf-Charms in Context.* The University of North Carolina Press, Chapel Hill, 1996.

Joralemon, Donald and Sharon, Douglas. *Sorcery and Shamanism: Curanderos and Clients in Northern Peru.* University of Utah Press, 1993.

Keikhefer, Richard. *Magic in the Middle Ages*. Cambridge University Press, 1989.

King, Francis. *The Secret Rituals of the O. T. O.* London: Daniel, 1973.

Kruskal, N. and R. Kobert. 'Agrostemma Githago.' *American Journal of Pharmacy and the Sciences*, Volume 22; Volume 64, 1892.

Lane, William Edward. *An Account of the Manners and Customs of Modern Egyptians*. London: Charles Knight and Co., 1846.

LaWall, Charles. *The Curious Lore of Drugs and Medicines: Four Thousand Years of Pharmacy.* Garden City Publishing Company, Graden City, NY 1927.

Lawson, J. Cuthbert. *Modern Greek Folklore and Ancient Greek Religion*. pp. 200–201.

Liddell, Henry George and Robert Scott. *A Greek-English Lexicon*. Oxford, Clarendon Press, 1940.

Logan, Patrick. *Irish Country Cures*. Appletree Press, 1981.

Lea, Henry Charles. *Materials Toward a History of Witchcraft* (Arthur C. Howland, ed.). Thomas Yoseloff, New York, NY 1957.

Leary, T. "Programmed Communication During Experiences With DMT." *The Psychedelic Review*, Issue 8, 1966.

MacMullen, Ramsay. *Christianity & Paganism in the Fourth to Eighth Centuries*. Yale University Press, 1981. Magnus, Albertus. (Best, Michael R.and Brightman, Frank H., eds.) *The Book of Secrets of Albertus Magnus*. Samuel Weiser, York Beach, Maine, 1999 (1550).

Mathers, S. L. MacGregor, trans. *The Book of the Sacred Magic of Abramelin the Mage*. Dover Publications, New York, 1975 (1458).

Moussaieff, Arieh et al. "Incensole acetate, an incense component, elicits psychoactivity by activating TRPV3 channels in the brain". Journal of the Federation of Experimental Biology, 22:3024–3034, 2008.

Merrifield, Ralph. *The Archaeology of Ritual and Magic*, New Amsterdam Books, 1987.

Mollenauer, Lynn Wood. *Strange Revelations: Magic, Poison, and Sacrilege in Louis XIV 's France*. The Pennsylvania State University Press, University Park, 2007.

National Trust, The. *Salt of the Earth: Origins and Meanings of Country Sayings*. Wiltshire, UK, 2002.

Nielsen, Kjeld. *Incense in Ancient Israel*. E. J. Brill, Leiden, 1986.

Ogden, Daniel. *Greek and Roman Necromancy*. Princeton University Press, 2001.

Ott, Jonathan. *Pharmacotheon: Entheogenic Drugs, Their Plant Sources, and History*. Natural Products Co., Kennewick, WA 1993.

Ayahuasca Analogues. Natural Products Co., Kennewick, WA 1994.

Patai, Raphael, *The Jewish Alchemists*. Princeton University Press, Princeton, NJ 1994.

Pendell, Dale. *Pharmako/Poeia: Plant Powers, Poisons, and Herbcraft*. Mercury House, 1995.

———. *Pharmako/Dynamis, 2002.*

———. *Pharmako/Gnosis, 2005.*

Pennacchio, Marcello et al. *Uses and Abuses of Plant-Derived Smoke: Its Ethnobotany as Hallucinogen*. Oxford, 2010.

Pennick, Nigel. *Secrets of East Anglian Magic*. Robert Hale, London, 1995.

Porter, Stephen. *The Great Plague*. Sutton Publishing Limited, Gloucestershire, UK,1999.

Rätsch, Dr. Christian. *A Dictionary of Sacred and Magical Plants*. Prism Press, Dorset, UK, 1992.

Plants of Love. Ten Speed Press, Berkeley, CA 1997.

Roper, Jonathan, ed. *Charms, Charmers, and Charming: International*

Research on Verbal Magic. Palgrave Macmillan, Basingstoke, 2009.

Ruland, Martin. *Lexicon of Alchemy* (trans A. E. Waite). Samuel Weiser, York Beach, ME 1984 (1612).

Schulke, D. "The Perfum'd Skull" *The Cauldron* 116, May 2005. *Viridarium Umbris*. Xoanon Limited, 2005

Ars Philtron: Codex Vasculum, Xoanon Limited 2008.

Lux Hæresis. Xoanon Limited, 2011. 175

Schultes, Richard Evans and Hoffman, Albert. *Plants of the Gods*.

Healing Arts Press, Rochester, VT 1992.

Semple, Gavin. "A Poisoned Chalice". Reineke Verlag, 2004.

Shah, Sayed Idries. *Oriental Magic*. Philosophical Library, Inc. New York, NY 1957.

The Secret Lore of Magic. Citadel Press, New York, 1970.

Sibly, Ebenezer. *A New and Complete Illustration of the Occult Sciences*,

Book 4, (1787).

Simmons, Marc. *Witchcraft in the Southwest: Spanish and Indian Supernaturalism on the Rio Grande*. University of Nebraska Press, Lincoln, 1974.

Simoons, Frederick J. *Plants of Life, Plants of Death*. University of Wisconsin Press, Madison, WI, 1998, p. 143.

Stary, Frantisek and Berger, Zdenek. *Poisonous Plants*. Magna Books, Leicester, UK (Aventinum, Prague) 1995.

Suleiman, Rami K. et al. "New withanolides from *Mandragora officinarum*: first report of withanolides from the Genus Mandragora", Department of Chemistry, University of Jordan, Amman, 2010.

Taube, Karl. *Aztec and Maya Myths*. University of Texas Press, Austin, TX, 1995.

Turkington, Carol. *Poisons and Antidotes*. Facts on File, NY, NY 1994.

Valentine, Basil. *The Triumphal Chariot of Antimony.* The Alchemical Press, 1992 (Amsterdam, 1685).

Wahshiya, Ibn. *The Book on Poisons.* (*Kitab al-sumum*) c. 950 AD; [trans. Levey, Martin. Philadelphia, PA 1966].

Weyer, Johann. *Witches, Devils, and Doctors in the Renaissance* (*De praestigiis daemonum et incantationibus ac veneficiis*), John Shea, trans. Tempe, AZ: Medieval and Renaissance Texts and Studies, Arizona State University, 1998 (1583).

White, David Gordon. *The Alchemical Body: Siddha Traditions in Medieval India.* University of Chicago Press, Chicago, 1996.

Wise, Michael, with Martin Abegg Jr. And Edward Cook, eds. *The Dead Sea Scrolls: A New Translation*. HarperCollins, San Francisco, 1996.

Wright, Elbee. *Book of Legendary Spells*. Marlar Publishing Co., Minneapolis, MN, 1968.

Younger, William. *Gods, Men and Wine*. World Publishing Company, Cleveland, 1966.

INDEX

This second edition of *Veneficium* was printed and bound on St Lucy's Eve, 2017 in an edition of three thousand copies, strictly comprising 2,200 softcover copies, 750 hardcover copies in black cloth and color wraps, and 50 hand numbered deluxe copies quarter bound in purple goatskin and slipcase.

Scribæ Quo Mysterium Famulatur